THE DESIGN OF
LOW-VOLTAGE, LOW-POWER
SIGMA-DELTA MODULATORS

THE KLUWER INTERNATIONAL SERIES IN ENGINEERING AND COMPUTER SCIENCE

ANALOG CIRCUITS AND SIGNAL PROCESSING
Consulting Editor: **Mohammed Ismail**. Ohio State University

Related Titles:

DISTORTION ANALYSIS OF ANALOG INTEGRATED CIRCUITS, Piet Wambacq, Willy Sansen; ISBN: 0-7923-8186-6
NEUROMORPHIC SYSTEMS ENGINEERING: Neural Networks in Silicon, edited by Tor Sverre Lande; ISBN: 0-7923-8158-0
DESIGN OF MODULATORS FOR OVERSAMPLED CONVERTERS, Feng Wang, Ramesh Harjani, ISBN: 0-7923-8063-0
SYMBOLIC ANALYSIS IN ANALOG INTEGRATED CIRCUIT DESIGN, Henrik Floberg, ISBN: 0-7923-9969-2
SWITCHED-CURRENT DESIGN AND IMPLEMENTATION OF OVERSAMPLING A/D CONVERTERS, Nianxiong Tan, ISBN: 0-7923-9963-3
CMOS WIRELESS TRANSCEIVER DESIGN, Jan Crols, Michiel Steyaert, ISBN: 0-7923-9960-9
DESIGN OF LOW-VOLTAGE, LOW-POWER OPERATIONAL AMPLIFIER CELLS, Ron Hogervorst, Johan H. Huijsing, ISBN: 0-7923-9781-9
VLSI-COMPATIBLE IMPLEMENTATIONS FOR ARTIFICIAL NEURAL NETWORKS, Sied Mehdi Fakhraie, Kenneth Carless Smith, ISBN: 0-7923-9825-4
CHARACTERIZATION METHODS FOR SUBMICRON MOSFETs, edited by Hisham Haddara, ISBN: 0-7923-9695-2
LOW-VOLTAGE LOW-POWER ANALOG INTEGRATED CIRCUITS, edited by Wouter Serdijn, ISBN: 0-7923-9608-1
INTEGRATED VIDEO-FREQUENCY CONTINUOUS-TIME FILTERS: High-Performance Realizations in BiCMOS, Scott D. Willingham, Ken Martin, ISBN: 0-7923-9595-6
FEED-FORWARD NEURAL NETWORKS: Vector Decomposition Analysis, Modelling and Analog Implementation, Anne-Johan Annema, ISBN: 0-7923-9567-0
FREQUENCY COMPENSATION TECHNIQUES LOW-POWER OPERATIONAL AMPLIFIERS, Ruud Easchauzier, Johan Huijsing, ISBN: 0-7923-9565-4
ANALOG SIGNAL GENERATION FOR BIST OF MIXED-SIGNAL INTEGRATED CIRCUITS, Gordon W. Roberts, Albert K. Lu, ISBN: 0-7923-9564-6
INTEGRATED FIBER-OPTIC RECEIVERS, Aaron Buchwald, Kenneth W. Martin, ISBN: 0-7923-9549-2
MODELING WITH AN ANALOG HARDWARE DESCRIPTION LANGUAGE, H. Alan Mantooth,Mike Fiegenbaum, ISBN: 0-7923-9516-6
LOW-VOLTAGE CMOS OPERATIONAL AMPLIFIERS: Theory, Design and Implementation, Satoshi Sakurai, Mohammed Ismail, ISBN: 0-7923-9507-7
ANALYSIS AND SYNTHESIS OF MOS TRANSLINEAR CIRCUITS, Remco J. Wiegerink, ISBN: 0-7923-9390-2
COMPUTER-AIDED DESIGN OF ANALOG CIRCUITS AND SYSTEMS, L. Richard Carley, Ronald S. Gyurcsik, ISBN: 0-7923-9351-1
HIGH-PERFORMANCE CMOS CONTINUOUS-TIME FILTERS, José Silva-Martínez, Michiel Steyaert, Willy Sansen, ISBN: 0-7923-9339-2
SYMBOLIC ANALYSIS OF ANALOG CIRCUITS: Techniques and Applications, Lawrence P. Huelsman, Georges G. E. Gielen, ISBN: 0-7923-9324-4
DESIGN OF LOW-VOLTAGE BIPOLAR OPERATIONAL AMPLIFIERS, M. Jeroen Fonderie, Johan H. Huijsing, ISBN: 0-7923-9317-1
STATISTICAL MODELING FOR COMPUTER-AIDED DESIGN OF MOS VLSI CIRCUITS, Christopher Michael, Mohammed Ismail, ISBN: 0-7923-9299-X
SELECTIVE LINEAR-PHASE SWITCHED-CAPACITOR AND DIGITAL FILTERS, Hussein Baher, ISBN: 0-7923-9298-1

THE DESIGN OF LOW-VOLTAGE, LOW-POWER SIGMA-DELTA MODULATORS

by

Shahriar Rabii
Level One Communications

Bruce A. Wooley
Stanford University

KLUWER ACADEMIC PUBLISHERS
Boston / Dordrecht / London

Distributors for North, Central and South America:
Kluwer Academic Publishers
101 Philip Drive
Assinippi Park
Norwell, Massachusetts 02061 USA
Telephone (781) 871-6600
Fax (781) 871-6528
E-Mail <kluwer@wkap.com>

Distributors for all other countries:
Kluwer Academic Publishers Group
Distribution Centre
Post Office Box 322
3300 AH Dordrecht, THE NETHERLANDS
Telephone 31 78 6392 392
Fax 31 78 6546 474
E-Mail <orderdept@wkap.nl>

 Electronic Services <http://www.wkap.nl>

Library of Congress Cataloging-in-Publication Data

A C.I.P. Catalogue record for this book is available
from the Library of Congress.

Copyright © 1999 by Kluwer Academic Publishers

All rights reserved. No part of this publication may be reproduced, stored in a retrieval system or transmitted in any form or by any means, mechanical, photo-copying, recording, or otherwise, without the prior written permission of the publisher, Kluwer Academic Publishers, 101 Philip Drive, Assinippi Park, Norwell, Massachusetts 02061

Printed on acid-free paper.

Printed in the United States of America

To Helena and Ariel

Contents

Contents vii

List of Figures xi

List of Tables xv

Preface xvii

1 Introduction 1
   1.1 Organization 3
   1.2 Simulation Details 4

2 Trends Toward Low-Voltage Power Supplies 7
   2.1 Technology Scaling in CMOS Processes 8
   2.2 Scaling Theory 9
   2.3 Battery Cell Technologies for Portable Electronic Systems 13

    2.3.1 Fundamentals of Battery Technology 13
    2.3.2 Discharge Characteristics 16
  2.4 Summary 18

3 Analog-to-Digital Conversion 21

  3.1 Analog-to-Digital Converters 21
    3.1.1 Quantizer Characteristics 23
    3.1.2 Performance Metrics 28
  3.2 Nyquist-Rate A/D Converters 29
  3.3 Oversampling A/D Converters 29
    3.3.1 Feedback Modulators 30
    3.3.2 First-Order Sigma-Delta Modulators 33
    3.3.3 Higher-Order Noise-Differencing Sigma-Delta Modulators 37
    3.3.4 Higher-Order Single-Stage Sigma-Delta Modulators 43
    3.3.5 Cascaded Sigma-Delta Modulators 43
    3.3.6 Multibit Sigma-Delta Modulators 49
  3.4 Summary 50

4 Power Dissipation in Sigma-Delta A/D Converters . 57

  4.1 Power Dissipation in a Sigma-Delta Modulator 57
  4.2 Ideal Integrator Power Dissipation 59
    4.2.1 Switched-Capacitor Integrator 59
    4.2.2 Continuous-Time Integrator 61
    4.2.3 Switched-Current Integrator 63
  4.3 Impact of Circuit Nonidealities 66
  4.4 Comparison of Amplifier Topologies 70
  4.5 Power Dissipation in the Decimation Filter 75
  4.6 Summary 76

5 Design of a Low-Voltage, High-Resolution Sigma-Delta Modulator 79

  5.1 Modulator Architecture 80
    5.1.1 Third-Order (2-1) Cascaded Modulator 80

		5.1.2 Fourth-Order Single Stage Modulator	84

      5.1.2 Fourth-Order Single Stage Modulator 84
      5.1.3 Second-Order Modulator with Multibit Quantization..........................85
  5.2 Signal Scaling 87
  5.3 Integrator Implementation 90
      5.3.1 Capacitor Matching 92
      5.3.2 Integrator Settling 93
  5.4 Circuit Noise 95
      5.4.1 Shaping of Circuit Noise 96
      5.4.2 Sampling Noise...97
      5.4.3 Amplifier Thermal Noise 103
      5.4.4 Flicker Noise 105
  5.5 Modulator Specifications..106
      5.5.1 Matching Requirements...106
      5.5.2 Settling Time 107
      5.5.3 Amplifier Gain Requirements 108
      5.5.4 Sampling Noise Budget 109
      5.5.5 Thermal Noise Budget 110
      5.5.6 Flicker Noise Attenuation..110
  5.6 Summary 111

6 Implementation of an Experimental Low-Voltage Modulator 115

  6.1 The Integrators 117
      6.1.1 The First Integrator..118
      6.1.2 The Second Integrator 120
      6.1.3 The Third Integrator 122
  6.2 The Operational Amplifiers..124
      6.2.1 First-Stage Operational Amplifier...128
      6.2.2 Second-Stage Operational Amplifier 131
      6.2.3 Third-Stage Operational Amplifier...132
  6.3 The Quantizers 133
      6.3.1 Comparators...133
      6.3.2 Digital-to-Analog Converters..135
  6.4 The Clocks..135
      6.4.1 Clock Generators...136
  6.5 Clock Boosters 137
      6.5.1 Voltage Doublers..138

6.6	Decimation Filtering	138
6.7	Experimental Results	139
6.8	Comparison of the Power Efficiency of A/D Converters	146
6.9	Summary	146

7 Conclusion 151

7.1 Recommendations for Further Investigation 152

A Fundamental Limits 155

A.1 Power in a Switched-Capacitor Integrator 155
A.2 Power in a Continuous-Time Integrator 158
A.3 Power in a Switched-Current Integrator 159

B Power Dissipation vs. Supply Voltage and Oversampling Ratio 163

B.1 Folded Cascode Amplifier 166
B.2 Two-Stage Class A Amplifier 168
B.3 Two-Stage Class A/AB Amplifier 170

C Effects of Capacitor Mismatch 173

D Test Setup 177

Index 185

List of Figures

Chapter 2
2.1 Cross-section of an N-well, CMOS transistor process. 8
2.2 Scaling parameters of an NMOS transistor. 9
2.3 Common-source feedback amplifier. 12
2.4 A NiCd battery cell in the charge and discharge phases. 14
2.5 Qualitative battery discharge curves. 17

Chapter 3
3.1 Block diagram of analog-to-digital conversion. 22
3.2 3-bit quantizer transfer curve. 24
3.3 Quantization error of a 3-bit quantizer. 25
3.4 Quantizer model using additive noise. 26
3.5 Quantizer error distribution under the white noise approximation. 26
3.6 Quantizer power spectral density under the white noise approximation. 27
3.7 Block diagram of a feedback modulator. 31
3.8 Linearized block diagram of a feedback modulator. 31
3.9 A first-order, 1-bit sigma-delta modulator. 33
3.10 Linearized model of a first-order sigma-delta modulator. 34
3.11 Block diagram manipulation of the second-order sigma-delta modulator. 39
3.12 Simplified block diagram of the second-order sigma-delta modulator. 40

3.13 Second-order sigma-delta modulator with integrator signal scaling............41
3.14 Block diagram of an Lth-order sigma-delta modulator................42
3.15 Block diagram of a 2-1 cascaded sigma-delta modulator. 44

Chapter 4
4.1 Sigma-delta modulator block diagram. 58
4.2 Switched-capacitor integrator................60
4.3 Continuous-time integrator................62
4.4 Switched-current integrator. 64
4.5 Operational amplifier with capacitive feedback................67
4.6 Amplifiers: (a) folded cascode, (b) two-stage class A, (c) two-stage class A/AB................71
4.7 First integrator power dissipation vs. supply voltage................74
4.8 First integrator power dissipation vs. oversampling ratio. 74

Chapter 5
5.1 Block diagram of a 2-1 cascaded sigma-delta modulator. 81
5.2 SNDR vs. input power in a second-order modulator................82
5.3 SNDR vs. input power in a 2-1 cascaded modulator................83
5.4 SNDR vs. input power showing signal dependence in noise floor................84
5.5 Spectral tones in a 2-1 cascaded modulator. 85
5.6 SNDR vs. input power in a fourth-order, single-stage modulator................86
5.7 SNDR vs. input power in a 4-bit, second-order modulator. 86
5.8 Probability density function of integrator outputs in the architecture of Figure 5.1 with $b = 2.5$, $\beta = 0.25$, and $\lambda = 1$................88
5.9 A 2-1 cascaded sigma-delta modulator with integrator gains. 89
5.10 Probability density function of integrator outputs in the architecture of Figure 5.9 with integrator gains shown in Table 5.1. 91
5.11 Simple switched-capacitor integrator. 92
5.12 (a) A switched-capacitor sampling network, (b) circuit model for sampling noise................97
5.13 Sampling noise during charge transfer in a switched-capacitor integrator................99
5.14 Sampling noise when using multiple sampling capacitors................101
5.15 Amplifier noise during the charge transfer phase................104
5.16 Simple switched-capacitor integrator with finite amplifier gain. 109

LIST OF FIGURES

Chapter 6

6.1 Block diagram of the experimental sigma-delta modulator. 116
6.2 Timing diagram of clock phases in the experimental modulator. 117
6.3 The first integrator without CDS. 118
6.4 The second integrator. 121
6.5 The third integrator. 123
6.6 A two-stage class A/AB operational amplifier. 125
6.7 (a) Differential-mode half circuit, (b) simplified differential-mode half circuit. 126
6.8 Common-mode half circuit 126
6.9 Common-mode feedback circuit for the first stage. 127
6.10 Common-mode feedback circuit for the second stage. 128
6.11 Biasing circuits. 129
6.12 Regenerative comparator. 134
6.13 Clock generator. 136
6.14 Clock boosting circuit. 137
6.15 Voltage doubler. 138
6.16 Die micrograph. 140
6.17 Measured SNR and SNDR vs. input power. 141
6.18 Measured baseband output spectrum. 142
6.19 Measured dynamic range vs. oversampling ratio. 142
6.20 Measured dynamic range and power dissipation vs. supply voltage..........144
6.21 Measured common-mode rejection ratio...144
6.22 Measured quantization tones. 145
6.23 Figure of merit vs. dynamic range of recent CMOS A/D converters.........147

Appendix D

D.1 Experimental test setup. 178
D.2 Differential sinewave generator termination circuit. 179
D.3 1.8-V power supply. 180
D.4 6-V voltage generator. 181
D.5 Bias voltage generator. 182
D.6 Bias current generator...182
D.7 Comparator circuit..183

List of Tables

Chapter 2
   2.1   Scaling strategies. 10
   2.2   Scaling guidelines. 11
   2.3   Key parameters of battery technologies. 18

Chapter 4
   4.1   Key parameters of three amplifier topologies. 72

Chapter 5
   5.1   Integrator gains. 91

Chapter 6
   6.1   Switches in the first integrator. 119
   6.2   Capacitor sizes in the first integrator. 119
   6.3   Switches in the second integrator. 120
   6.4   Capacitor sizes in the second integrator. 121
   6.5   Switches in the third integrator. 124
   6.6   Capacitor sizes in the third integrator. 124
   6.7   Transistor sizes in the first amplifier. 130
   6.8   Capacitor sizes in the first amplifier. 130

6.9 Transistor sizes in the second amplifier. 131
6.10 Capacitor sizes in the second amplifier. 132
6.11 Transistor sizes in the third amplifier. 132
6.12 Capacitor sizes in the third amplifier. 133
6.13 Transistor sizes in the regenerative latches. 134
6.14 Performance summary. 143

Appendix D
D.1 Test setup equipment list. 179

Preface

Two factors, the continued scaling of device dimensions and the explosive growth in the demand for portable electronic systems, are forcing a significant reduction in the supply voltage from which circuits integrated in VLSI technologies operate. While lowering the supply voltage affords dramatic power savings in most digital circuits, it severely complicates the design of high-resolution analog circuits. For example, the power dissipated in many analog circuits actually increases as the supply voltage is reduced. This monograph, which is based on the first author's doctoral dissertation, explores the implementation of precision low-voltage analog circuits within the context of designing a sigma-delta ($\Sigma\Delta$) modulator capable of digitizing audio-band signals.

Oversampling techniques based on $\Sigma\Delta$ modulation are widely used to implement the interfaces between analog and digital signals in VLSI systems. This approach is relatively insensitive to imperfections in circuit components and offers numerous advantages for the realization of high-resolution analog-to-digital (A/D) converters in the low-voltage environment that is increasingly demanded by advanced VLSI technologies and by portable systems. In particular, oversampling architectures are a potentially power-efficient means of implementing high-resolution A/D converters because they reduce the number and complexity of the analog circuits in comparison with Nyquist-rate converters. Furthermore, they allow the performance requirements, and thus most of the power dissipation, to be concentrated in the input stage of a converter; specifically, the power dissipation of a highly oversampled $\Sigma\Delta$ modulator is shown herein to be fundamentally limited by that of a single integrator with the resolution and bandwidth required for a specified application.

The reduction of the supply voltage presents a number of significant challenges to the implementation of analog circuits with high dynamic range. The issues of output swing, common-mode rejection, and power-supply rejection play an increasingly important role in determining the performance that can be achieved. Insight gained from the theoretical analysis of modulator and integrator power dissipation has been used in this work to design an experimental low-voltage A/D converter with digital-audio performance. The converter has been implemented in a conventional 0.8-μm CMOS technology with device thresholds on the order of 700 mV. When operated from a 1.8-V supply, it achieves a dynamic range of 99 dB at a signal bandwidth of 25 kHz while dissipating only 2.5 mW.

Following a brief introduction, this text begins with an overview of the two factors driving the reduction in the supply voltage for VLSI systems – technology scaling and battery life in portable systems. Metrics by which the performance of an A/D converter is measured and the basic principles of oversampling A/D conversion are then reviewed in Chapter 3. Chapter 4 explores the fundamental limits on power dissipation in $\Sigma\Delta$ modulators. The architecture and circuit design specifications for a low-voltage, low-power $\Sigma\Delta$ modulator are established in Chapter 5, and Chapter 6 presents the circuit design details and measured performance for an experimental modulator.

The authors are very much indebted to numerous colleagues among the students and staff at Stanford, as well as at other organizations and institutions, for their many contributions to the work upon which the research leading to this monograph is based.

Shahriar Rabii

Bruce A. Wooley

CHAPTER 1 *Introduction*

Rapid growth in the demand for portable, battery-operated electronics for communications, computing and consumer applications, as well as the continued scaling of VLSI technology, has begun to significantly alter the constraints under which many semiconductor integrated circuits are designed. In particular, in order to both conserve power in digital circuits and reduce the high electric fields that accompany the scaling of device dimensions, it is becoming necessary for circuits to operate from reduced supply voltages. Without the use of voltage regulation, the minimum supply voltage in portable equipment is generally the end-of-life battery voltage multiplied by the number of cells connected in series. In the case of nickel-cadmium and alkaline cells, the end-of-life voltage is 0.9 V, corresponding to a 1.8-V minimum supply voltage for two batteries in series.

Considerations such as cost, performance and portability are powerful incentives for integrating analog and mixed-signal circuit functions, such as data conversion, on the same chip as large digital data and signal processing circuits. However, even supply voltages as large as 5 V severely constrain the dynamic range available for realizing analog circuits with the performance that is increasingly demanded in communications and multimedia applications. Thus, it is an especially challenging task to maintain the desired levels of performance as the supply voltage is lowered. Moreover, while a reduction in supply voltage generally results in significant power savings in digital circuits, the power consumed in analog circuits is actually likely to increase.

Introduction

Among the most critical, and often performance limiting, functions in mixed-signal VLSI circuits are the interfaces between analog and digital representations of information. As a consequence, considerable attention is being given to the design of CMOS analog-to-digital converters that operate from supply voltages below 5 V with reduced power dissipation. Much of the work reported to date focuses on the realization of low-power, low-voltage oversampling converters for voiceband telephone applications [1.1]-[1.5] and the use of pipelining and folding to implement power-efficient video-rate converters [1.6]-[1.9]. In the case of oversampling converters, alternative CMOS circuit approaches, such as switched-capacitor, switched-current and continuous-time circuits, have been explored.

Sigma-delta ($\Sigma\Delta$) modulation is a robust means of implementing high-resolution analog-to-digital converters in a VLSI technology. By combining oversampling and feedback to shape the noise, and then using a digital lowpass filter to attenuate the noise that has been pushed out-of-band, it is possible to achieve a dynamic range as high as 16 bits or more at relatively modest oversampling ratios [1.10]. Moreover, as shown herein, oversampling architectures are potentially a power-efficient means of implementing high-resolution A/D converters. In effect, an increase in sampling rate can be used to reduce the number and complexity of the analog circuits required in comparison with Nyquist-rate architectures, transferring much of the signal processing into the digital domain where power consumption can be dramatically reduced simply by scaling the technology and reducing the supply voltage.

This text describes the results of research into the design of oversampling sigma-delta modulators that provide the performance required for high-fidelity, digital-audio applications when operated from supply voltages as low as 1.5 V to 1.8 V and dissipating only a few mW of power [1.11], [1.12]. The objective of this work has been to demonstrate a modulator architecture and constituent circuits that can deliver a high dynamic range over a 25-kHz bandwidth and can be fabricated in a standard CMOS technology.

An experimental modulator described herein employs switched-capacitor circuits and has been integrated in a 0.8-μm CMOS technology with NMOS and PMOS threshold voltages of +0.65-V and –0.75-V, respectively. It provides a dynamic range of 99 dB and a peak signal to noise plus distortion ratio of 95 dB across a signal bandwidth of 25 kHz. It operates over a supply range of 1.5-V to 2.5-V and dissipates 2.5 mW at 1.8 V.

1.1 Organization

Chapter 2 outlines the motivation behind the drive towards low voltage operation in mixed-signal circuits. First, the scaling of VLSI semiconductor technologies and its impact on the power supply voltage are considered. The current state of the art and predictions for the coming years are examined. Then follows an introduction to battery technology, an important consideration in portable system design. After a summary of the principles of operation, an overview of dominant and emerging battery technologies is presented.

Chapter 3 begins with a consideration of quantized signals in which the relationships among quantization noise, the input signal, and the quantizer resolution are explored. After a brief description of Nyquist rate A/D converters, the principles of oversampling and feedback in the context of A/D conversion are introduced. Architectures for a subclass of oversampling converters, sigma-delta modulators, are considered, and their performance is evaluated.

In Chapter 4, an analysis of power dissipation in $\Sigma\Delta$ modulators is presented. After considering issues in low-power modulator design, a detailed analysis of the dominant power dissipating element, the first integrator, is given. The fundamental limits on power dissipation for several integrator implementations are examined. The analysis is extended to include the impact of practical circuit non-idealities and then focuses on a comparison of amplifier topologies for the realization of low-voltage, low-power integrators. The chapter closes with a brief review of power dissipation in the digital decimation filters that are needed to convert the low-resolution, high-speed digital output of a $\Sigma\Delta$ modulator into a high-resolution, lower speed digital encoding.

In Chapter 5, various $\Sigma\Delta$-modulator architectures are considered, and a topology is identified that provides a favorable set of design trade-offs for low-voltage, low-power operation. The limitations of this architecture are evaluated via both hand analyses and simulations. Behavioral level simulations are then used to aid in choosing integrator gains so as to scale the internal signal ranges and optimize the modulator overload characteristics. Following an analysis of noise sources, circuit specifications are established for an experimental modulator.

Chapter 6 presents a detailed description of the circuits used to implement the modulator described in Chapter 5. Experimental results for a prototype fabricated in a 0.8-μm CMOS process are presented. Based on the analysis in Chapter 4, a figure of merit for the power dissipation of A/D converters is proposed, and the power efficiency of some recently reported designs is compared.

Introduction

The results of this work are summarized in Chapter 7, and additional areas of research are suggested.

1.2 Simulation Details

The simulation results presented in this work were generated using the program MIDAS[1] [1.13], [1.14]. Spectra were estimated using a fast Fourier transform (FFT) in conjunction with a windowing function described by Nuttall [1.15]. Decimation filtering was performed using MIDAS. The decimation filter architecture comprises a cascade of sinc filters followed by an FIR filter. The FIR filter coefficients were generated with the program Matlab [1.16] using the frequency sampling method.

REFERENCES

[1.1]   E. J. van der Zwan and E. C. Dijkmans, "A 0.2mW CMOS ΣΔ modulator for speech coding with 80dB dynamic range," *ISSCC Digest of Tech. Papers*, pp. 232-233, February 1996.

[1.2]   S. Kiriaki, "A 0.25mW sigma-delta modulator for voice-band applications," *Symp. on VLSI Circuits Digest of Tech. Papers*, pp. 35-36, 1995.

[1.3]   N. Tan and S. Eriksson, "A low-voltage switched-current delta-sigma modulator," *IEEE J. Solid-State Circuits,* vol. SC-30, pp. 599-603, May 1995.

[1.4]   S. S. Bazarjani, M. Snelgrove, and T. MacElwee, "A 1V switched-capacitor ΣΔ modulator," *Symp. on Low Power Electronics Digest of Tech. Papers*, pp. 70-71, October 1995.

[1.5]   J. Grilo, E. MacRobbie, R. Halim, and G. Temes, "A 1.8V 94dB dynamic range ΣΔ modulator for voice applications," *ISSCC Digest of Tech. Papers*, pp. 232-233, February 1996.

---

1. More information can be obtained about MIDAS at the following location in the World Wide Web: http://cis.stanford.edu/icl/wooley-grp/midas.html

[1.6]   K. Kusumoto, A. Matsuzawa, and K. Murata, "A 10-b 20-MHz 30-mW pipelined interpolating CMOS ADC," *IEEE J. Solid-State Circuits,* vol. SC-28, pp. 1200-1206, December 1993.

[1.7]   M. Yotsuyanagi, H. Hasegawa, M. Yamaguchi, M. Ishida, and K. Sone, "A 2V 10b 20MSample/s mixed-mode subranging CMOS A/D converter," *ISSCC Digest of Tech. Papers,* pp. 282-283, February 1995.

[1.8]   T. B. Cho and P. R. Gray, "A 10 b, 20 Msample/s, 35 mW pipeline A/D converter," *IEEE J. Solid-State Circuits,* vol. SC-30, pp. 166-172, March 1995.

[1.9]   A. G. W. Venes and R. J. van de Plassche, "An 80MHz 80mW 8b CMOS folding A/D converter with distributed T/H preprocessing," *ISSCC Digest of Tech. Papers,* pp. 318-319, February 1996.

[1.10] J. Candy and G. Temes, "Oversampling methods for A/D and D/A conversion," in *Oversampling Delta-Sigma Data Converters,* pp. 1-29, New York: IEEE Press, 1992.

[1.11] S. Rabii and B. A. Wooley, "A 1.8V, 5.4mW, digital-audio $\Sigma\Delta$ modulator in 1.2μm CMOS," *ISSCC Digest of Tech. Papers,* pp. 228-229, February 1996.

[1.12] S. Rabii and B. A. Wooley, "A 1.8V digital-audio $\Sigma\Delta$ modulator in 0.8μm CMOS," *IEEE J. Solid-State Circuits,* vol. SC-32, pp. 783-796, June 1997.

[1.13] L. Williams, and B. A. Wooley, "MIDAS – a functional simulator for mixed digital and analog sampled data systems," *Proc. 1992 IEEE Int. Symp. Circuits Syst.,* pp. 2148-2151, May 1992.

[1.14] S. Rabii, L. Williams, B. Boser, and B. A. Wooley, *MIDAS User Guide Version 3.1,* Stanford University, Stanford, CA, 1997.

[1.15] A. Nuttall, "Some windows with very good sidelobe behavior," *IEEE Trans. on Acoustics, Speech, and Signal Processing,* vol. ASSP-29, pp. 84-89, February 1981.

[1.16] The MathWorks, *Matlab User Guide,* The MathWorks, Inc., Natick, MA, 1994.

CHAPTER 2
Trends Toward Low-Voltage Power Supplies

Two factors, technology scaling and battery life in portable electronics, are accelerating the reduction in the supply voltage used for CMOS VLSI circuits beyond what might be expected from historical trends. While the scaling of transistor dimensions results in dramatic increases in both the speed and density of digital circuits, it also results in a corresponding increase in the electric fields within the device if the supply voltage is held constant. Breakdown considerations thus make it increasingly difficult to sustain the constant supply voltage that has characterized the past two decades of advances in CMOS technology. Moreover, lowering the supply voltage significantly reduces the energy consumed per operation in a digital system.

Explosive growth in the market for portable, battery operated electronics has stimulated the demand for low-power circuits that are capable of operating from a low supply voltage. Battery characteristics such as operating voltage and discharge behavior vary greatly with the particular battery chemistry and construction, and can impose widely differing requirements on the circuits they supply. This chapter examines both VLSI technology scaling and battery technology in order to highlight the relevance of the research presented in this monograph to some of the important issues faced by the electronics industry.

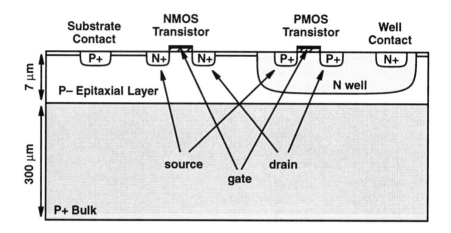

FIGURE 2.1 Cross-section of an N-well, CMOS transistor process.

2.1 Technology Scaling in CMOS Processes

CMOS is the dominant technology for manufacturing integrated electronic circuits. Shown in Figure 2.1 is a simplified cross-section of an NMOS and a PMOS transistor as implemented in an epitaxial, N-well, CMOS process. The source and drain terminal characteristics of the device are controlled by modulating the number of carriers in the channel with the gate voltage [2.1]. The key to increasing the computational power of a digital integrated circuit is simply to increase the number of transistors on a single die by reducing the size of the devices and increasing the area of the chip. The result of such scaling has been an exponential growth in computational capacity. For example, a state-of-the-art microprocessor has been integrated on a 314-mm^2 die with 15.2 million transistors of 0.25-µm gate length, operates at a 600-MHz clock rate and dissipates 72W from a 2-V supply [2.2].

Unfortunately, the increase in both the number of digital gates per chip and the speed of those gates that accompanies the scaling of technology has resulted in significant increases in power dissipation. Lowering the supply voltage is an obvious and straightfoward means of reducing the power dissipated in digital circuits, thereby relaxing the cooling requirements and extending the battery life of portable electronic systems.

Scaling Theory

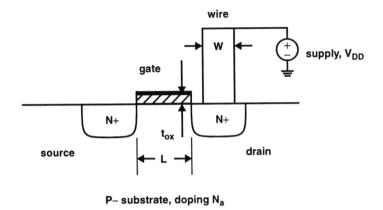

FIGURE 2.2 Scaling parameters of an NMOS transistor.

Cost and performance considerations provide powerful incentives for integrating the interfaces between the analog and digital representations of information on the same chip as the digital signal and data processing. There is thus a growing demand for analog circuits that can deliver high performance in a low-voltage, VLSI technology designed primarily to meet the needs of digital circuits. Unfortunately, the reduction in supply voltage that seems likely to accompany the continued scaling of VLSI technologies significantly complicates the design of analog circuits.

2.2 Scaling Theory

This section provides a brief introduction to CMOS scaling. While a thorough discussion of the subtle trade-offs inherent in CMOS scaling is beyond the scope of this work, more detailed information can be obtained from the large body of published work, including references [2.3]-[2.6].

The transistor and interconnect parameters shown in Figure 2.2 can be scaled in a variety of ways relative to one another. Basic algorithms employed to scale CMOS technology include *constant field*, *constant voltage*, and *generalized* scaling [2.4]-[2.6]. Table 2.1 summarizes the scaling relationships among the parameters in Figure 2.2 for each of these approaches.

TABLE 2.1 Scaling strategies.

Parameter	Constant Field	Constant Voltage	Generalized scaling
Dimensions (L, W, t_{ox}, ...)	$1/\alpha$	$1/\alpha$	$1/\alpha$
Substrate doping	α	α^2	$\varepsilon\alpha$
Supply voltage	$1/\alpha$	1	ε/α
Electric field	1	α	ε
Gate delay	$1/\alpha$	$1/\alpha^2$	$1/(\varepsilon\alpha)$
Power per function	$1/\alpha^2$	α	ε^3/α^2
Power per unit area	1	α^3	ε^3

In constant field scaling, a dimensionless scaling factor, α, is applied such that the electric field intensity and pattern remain constant as the device dimensions are scaled. Important consequences of constant field scaling are a decrease in the device area of α^2, a decrease in propagation delay of α, a decrease in the power per function of α^2, and a constant power per unit area. The practical limitations of constant field scaling arise from transistor parameters that cannot be scaled effectively. For example, lowering the supply voltage must typically be accompanied by a reduction in transistor threshold voltages, which results in an increase in subthreshold leakage currents. However, in many systems constraints on the allowable standby current impose a lower limit on the threshold voltages and, hence, the supply voltage. In addition, strict constant field scaling would require a change in supply voltage with each generation of technology. From a system standpoint, this is generally not practical since systems are likely to include components implemented in different generations of technology. Finally, higher performance can often be achieved by allowing electric fields to increase in scaled devices.

In constant voltage scaling, the supply voltage is kept constant while the device dimensions are reduced. This results in an increase in the electric fields within the device, leading to a dramatic improvement of α^2 in speed and an α increase in power per function. Therefore, the power per unit area increases at a rate of α^3. The problems associated with this scaling approach include the deleterious effects of increased electric fields on device performance and reliability, as well as a rapid increase in power dissipation. Historically, breakdown limitations have been addressed through extensive drain engineering in MOS devices [2.7], [2.8]. However, at submicron dimensions electric fields approach the maximum sustainable levels, necessitating some reduction in supply voltage.

Scaling Theory

Generalized scaling refers to a combination of constant field and constant voltage scaling. By introducing a second scaling factor, ε, a compromise between power and performance can be achieved. In fact, a scaled process can be optimized either for speed or power. Thus, two processes with identical feature sizes may be designed to accommodate different supply levels through the use of different doping profiles and oxide thicknesses. In all scaling scenarios, however, the power supply voltage must eventually be reduced. The scaling guidelines shown in Table 2.2 [2.6] represent one view of the direction that the electronics industry may take in the coming years.

TABLE 2.2 Scaling guidelines.

	1995	1998	2001	2004
Supply (V) (High Perf.)	3.3/2.5	2.5/1.8	1.5	1.2
Supply (V) (Low Power)	2.5/1.5	1.5/1.2	1.0	1.0
Channel Length (μm)	0.35	0.2	0.1	0.07
Oxide Thickness (nm)	9	6	3.5	2.5
Relative Density	6.3	12.8	25	48
Rel. Speed (High Perf.)	2.7/3.4	4.2/5.1	7.2	9.6
Rel. Speed (Low Power)	2.0/2.4	3.2/3.5	4.5	7.2
Rel. Power/Function (High Perf.)	0.5/0.3	0.3/0.2	0.1	0.08
Rel. Power/Function (Low Power)	0.20/0.09	0.08/0.06	0.04	0.04
Rel. Power/Unit Area (High Perf.)	3.0/2.1	3.7/2.3	3.1	3.7
Rel. Power/Unit Area (Low Power)	1.3/0.6	1.0/0.7	0.9	2.0

From an observation of historical trends and recent research results, it seems unlikely that supply voltages will be scaled as aggressively as suggested by Table 2.2. When considering the impact of process and temperature variations on the energy delay product, the highest energy efficiency can be obtained with a supply voltage of about 1.5 V [2.9]. Furthermore, there are indications that as the gate oxide thickness is reduced to 4 nm and lower, the number of defects in the oxide drop significantly, allowing the gate to support a higher electric field [2.10].

While the scaling of transistors and the associated reduction in supply voltage are highly desirable for digital circuits, they pose some important difficulties in analog and mixed-signal circuit design. Prominent among these is a reduction in signal swing that can limit the achievable dynamic range and increase the harmonic distortion

Trends Toward Low-Voltage Power Supplies

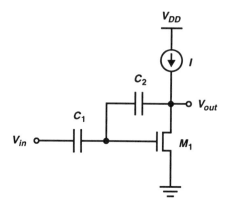

FIGURE 2.3 Common-source feedback amplifier.

introduced by analog circuits. For example, in the common-source feedback amplifier shown in Figure 2.3, the linear output swing is limited by the variation in gain that results from change in the output conductance of M_1 as a function of the output voltage. As the drain-to-source voltage approaches the saturation voltage of the transistor, the output conductance increases rapidly, reducing the open loop gain. Gain variation together with low open loop gain give rise to significant harmonic distortion in the response of the amplifier.

In addition to the lower supply voltages, other aspects of technology scaling present mixed-signal design challenges. In velocity saturated devices, the transistor transconductance increases very slowly as the gate length is reduced. However, the output conductance rises rapidly, resulting in a precipitous drop in the intrinsic gain of a transistor. If the transistor gain is held constant by increasing the channel length beyond minimum, there is unlikely to be a significant bandwidth improvement in a scaled technology. Another issue is the increased thermal noise of scaled devices due to hot electrons whose noise temperature is higher than the lattice temperature [2.11]. In addition, the lower flicker noise characteristic of PMOS transistors in comparison with NMOS transistors will probably vanish since the PMOS transistor current will flow at the silicon-silicon dioxide interface rather than deeper in the channel, as was the case in older technologies. Transistor matching, which is related to the variation in the number of doping atoms in the channel depletion layer, will improve for a given gate area as a process is scaled [2.12]. However, the improvement in matching will be modest and will not keep pace with the reduction in the signal swings as the supply voltage lowered. Thus, mismatch-induced offsets will be an ever increasing fraction of the signal levels, making the design of offset sensitive circuits more problematic. Finally, passive components such as capacitors and resistors may not be available in aggressively scaled standard digital technologies. The bottom plate parasitics of the capacitors and the voltage coefficient and process variability of the resistors that can be implemented in standard technologies may thus introduce additional problems for the mixed-signal circuit designer.

As highlighted above, analog and mixed-signal design in aggressively scaled digital processes is likely to become an increasingly difficult challenge. Although improvements in circuit design and system architecture will address many of these issues, it remains unclear as to whether it will be possible to achieve the level of performance required in future applications with a single, mixed-signal IC. An alternative is the development of technologies that specifically address some of the needs of analog circuit design. For example, technologies might be developed that offer options of thicker oxides, multiple threshold voltages, and high-quality passive components in the same process as fine geometry digital transistors and interconnects.

2.3 Battery Cell Technologies for Portable Electronic Systems

The battery life, weight, and volume needed to supply energy to a portable electronic system are typically dominant considerations in the design of that system. These parameters are closely related to the characteristics of the battery technology used to power the system. This section presents an overview of battery technology and summarizes the current state-of-the-art.

2.3.1 Fundamentals of Battery Technology

A battery is composed of one or more *battery cells* that are connected either in parallel or in series. A battery cell is a device that converts chemical energy into electrical energy through a reduction-oxidation (redox) process [2.13]. Shown in Figure 2.4 is a conceptual illustration of a NiCd battery cell. It comprises an anode electrode where oxidation occurs, a cathode electrode where reduction occurs, and an electrolyte that is ionically conductive but electrically insulating. In practical cells there is normally a separator between the electrodes to insure that an internal electrical short circuit does not occur.

Trends Toward Low-Voltage Power Supplies

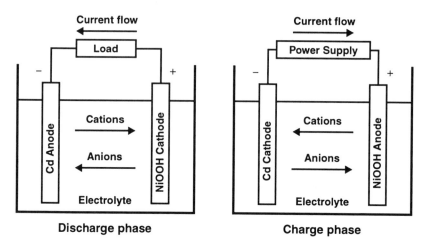

FIGURE 2.4 A NiCd battery cell in the charge and discharge phases.

The key idea behind the operation of a battery cell is that the electrolyte allows the flow of ions within the battery but prevents the flow of electrons. For the reaction to proceed, electrons must flow between the electrodes through an external circuit. Battery cells normally operate in the discharge phase as electrons flow from the negative electrode, the anode, to the positive electrode, the cathode, through an external load. In some batteries, the electrochemical reaction is reversible and the chemical energy of the cell can be restored by operating in the charge phase wherein the anode becomes the positive electrode and the cathode becomes the negative electrode and electrons are supplied by an external power source. Batteries that can only operate in the discharge phase are known as *primary* batteries, while rechargeable ones are known as *secondary* batteries.

A nickel-cadmium cell is a secondary battery. The discharge reaction at the cadmium anode of a nickel-cadmium cell into cadmium hydroxide is described by

$$Cd + 2OH^- \rightarrow Cd(OH)_2 + 2e^-. \tag{2.1}$$

At the cathode, nickel oxyhydroxide is reduced to nickel hydroxide by the following reaction

$$NiOOH + H_2O + e^- \rightarrow OH^- + Ni(OH)_2. \tag{2.2}$$

Battery Cell Technologies for Portable Electronic Systems

These reactions can be reversed by supplying electrons to the cell.

The theoretical maximum voltage of a cell is the sum of the anode oxidation potential and the cathode reduction potential. In the case of nickel-cadmium cells, the anode oxidation potential is 0.49 V and the cathode reduction potential is 0.81 V, resulting in a theoretical cell voltage of 1.3 V. The theoretical battery cell specific capacity is the charge that a cell can deliver to a load per gram of electrode material, which is known as the *gram-equivalent* weight of the cell. The theoretical capacity density limit is the charge that can be supplied by 1 gram of electrons, which is 96,487 coulombs (C) or 26.8 ampere-hours (Ah). In the case of the nickel-cadmium cell, the theoretical specific capacity is 0.18 Ah/g.

The cell voltage and capacity observed in practice are substantially lower than the theoretical predictions and vary greatly with the discharge and environmental conditions. This loss is a consequence of three factors: resistive losses, activation polarization, and concentration polarization. The resistive losses are due to the ohmic resistance of the electrodes, the ionic resistance of the electrolyte, and the ohmic resistance of the battery housing that connects the electrodes to the load. Although this loss mechanism has a number of distinct physical sources, it is well modeled by a linear resistor in series with the cell. Activation polarization refers to the potential required to initiate the electrochemical reaction at the surface of the electrodes. It is a non-linear phenomenon that depends strongly on the chemistry of the cell. Concentration polarization refers to the local depletion of ions in the electrolyte in the vicinity of an electrode. Mass transfer by diffusion then becomes a limiting factor and produces an appreciable voltage drop across the depleted layer. This loss mechanism is non-linear, rising quickly at low current levels and then saturating when the maximum diffusion current of the electrolyte is reached [2.13], [2.14]. The effects of both activation polarization and concentration polarization can be mitigated by increasing the area of the electrodes. Many manufacturers use porous electrodes to achieve high surface areas without increasing the cell size.

A common method of specifying practical cell capacity, which accounts for the above loss mechanisms, is the *C* rate. In this measure, the number following the letter '*C*' is the amount of time, expressed in hours, obtained by dividing the total charge that a battery cell can deliver to a constant current load by the load current. For example, a battery that has a $C/10$ capacity of 2 Ah can deliver 200 mA for 10 hours. Due to the nonlinearities of the discharge curve, the *C* rate scale is also nonlinear. Typically, discharging at a low *C* rate yields higher battery capacity than discharging at a high rate. Thus, a battery with a $C/10$ capacity of 2 Ah is likely to have a $C/5$ capacity of less than 2 Ah and a $C/20$ capacity of more than 2 Ah.

Trends Toward Low-Voltage Power Supplies

With constant power discharge cycles becoming more popular, particularly for portable electronic equipment, an alternative measure called the E rating system has come into use. In this scale, the number following the letter 'E' is the amount of time, expressed in hours, obtained by dividing the total energy that a battery cell can deliver to a constant power load by the power dissipated in the load. For example, a battery that has an $E/20$ capacity of 1.2 Wh can supply 60 mW for 20 hours. Like the C rate scale, the E rate scale is nonlinear.

Desirable properties for electrodes are high redox potentials, high capacity, low volume, good conductivity, chemical and thermal stability, large surface area, low toxicity, and low cost. Important issues in electrolytes are ionic conductivity, electrical conductivity, reactivity with the electrodes, chemical and thermal stability, toxicity, and cost.

2.3.2 Discharge Characteristics

The actual amount of energy delivered by a battery cell during a discharge cycle depends strongly on the discharge conditions. Important factors are whether the load is pulsed or continuous and, in the case of continuous loads, whether it is a constant current or constant power load. Figure 2.5 illustrates qualitatively the discharge profile of batteries under these loading conditions assuming the same power at the end voltage. Pulsed loads generally result in greater battery capacity. This is due to a reduction in the deleterious effects of concentration polarization, since the electrolyte ionic concentration can be restored when the load is disconnected. For a significant increase in capacity, the switching period of the load must be longer than the diffusion time of ions in the electrolyte, yet short enough so that significant concentration polarization does not occur.

When discharged through a continuous load, constant-current discharge corresponds to a system that operates directly from a battery without a voltage regulator and draws a supply-independent current. Since the operating voltage of the battery falls throughout its discharge cycle, such a system should be designed to operate with a supply voltage as low as the end-of-life voltage of the battery in order to utilize the cells effectively.

A constant-power load corresponds to a system that uses a voltage regulator and consumes a constant current from a fixed voltage supply. In general, constant-power loads extract more energy from the battery than constant-current loads and achieve a longer life span. However, constant-power loads introduce the additional power dissi-

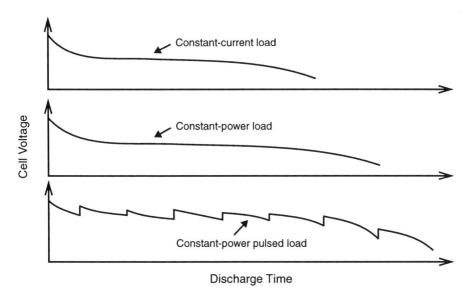

FIGURE 2.5 Qualitative battery discharge curves.

pation of the regulator, which may increase their total power dissipation compared to the constant current load.

A significant environmental factor in the performance of batteries is the ambient temperature. Low temperatures reduce the rate of chemical processes within the cell and increase internal resistances, thereby degrading capacity. At high temperatures, chemical breakdown can occur within the cell, also reducing capacity. The optimum temperature for maximum battery performance depends on the type of cell materials and construction and is normally between 20°C and 40°C.

Other issues in selecting a battery technology for a particular application are the degradation of the battery before it is put into use, known as the shelf life; the variation of the battery voltage across its life time, known as the discharge profile; and in the case of secondary batteries, the number of times it can be recharged, or its cycle life. There are currently many different battery technologies, each with its own distinct advantages. Table 2.3 summarizes important characteristics of several primary and secondary batteries. The dominant primary battery technologies are zinc-carbon and alkaline. Recently, lithium anode cells have attracted a great deal attention and interest because of their high capacity and voltage. Among secondary batteries, the nickel-

TABLE 2.3 Key parameters of battery technologies.

	Primary battery technologies			Secondary battery technologies		
	Zinc-carbon	Alkaline	Zinc-air	Nickel-cadmium	Nickel-metal-hydride	Lithium ion
Nominal Voltage (V)	1.5	1.5	1.5	1.2	1.2	4.0
Operating Voltage (V)	1.2	1.2	1.3	1.1	1.1	3.0
End-of-Life Voltage (V)	0.9	0.9	0.9	1.0	1.0	2.5
Energy Density (Wh/kg)	65	125	340	35	50	90
Energy Density (Wh/L)	100	330	1050	100	175	200
Cost	low	moderate	moderate	moderate	high	high

cadmium chemistry dominates consumer portable electronics applications. Alternative secondary batteries are nickel-metal-hydride and lithium-ion. Although the cost of the alternative battery technologies remains high, continued development efforts may soon make them competitive with the current industry standards [2.13], [2.14].

2.4 Summary

In this chapter CMOS technology scaling and battery operation have been considered in the context of low-voltage and low-power circuits and systems. The aggressive scaling of CMOS technologies has provided for an exponential growth in the computational power obtainable in a single integrated circuit. However, the accompanying scaling of the supply voltage and changes in transistor characteristics present significant challenges to the continued integration of mixed analog-digital functions on the same die as large digital circuits.

Summary

The growth in the demand for portable electronics also imposes increasingly severe voltage and power dissipation constraints on the design of mixed-signal circuits. Low-voltage operation reduces the number of batteries that must be connected in series to reach the required potential, while low power dissipation reduces the volume and weight of batteries needed to achieve a particular lifetime.

REFERENCES

[2.1] R. Muller and T. Kamins, *Device Electronics for Integrated Circuits*, John Wiley & Sons, 1986.

[2.2] B. A. Gieseke, et al., "A 600MHz superscalar RISC microprocessor with out-of-order execution," *ISSCC Digest of Tech. Papers*, pp. 176-177, February 1997.

[2.3] J. D. Meindl, "Low power microelectronics: retrospect and prospect," *Proceedings of the IEEE*, vol. 83, no. 4, pp. 619-35, April 1995.

[2.4] R. Dennard, F. Gaensslen, H. Yu, V. Rideout, E. Bassous, and A. LeBlanc, "Design of ion-implanted MOSFET's with very small physical dimensions," *IEEE J. Solid-State Circuits*, vol. SC-9, pp. 256-267, October 1974.

[2.5] G. Baccarani, M. Wordeman, and R. Dennard, "Generalized scaling theory and its application to 1/4 micron MOSFET design," *IEEE Trans. on Electron Devices*, vol. ED-31, pp. 452-462, April 1984.

[2.6] B. Davari, R. Dennard, and G. Shahidi, "CMOS scaling for high performance and low power – the next ten years," *Proceedings of the IEEE*, vol. 83, pp. 595-606, April 1995.

[2.7] K. W. Terrill, C. Hu, and P. K. Ko, "An analytical model for the channel electric field in MOSFETs with graded-drain structures," *IEEE Electron Device Letters*, vol. EDL-5, no. 11, pp. 440-2, November 1984.

[2.8] L. Su, et al., "A high-performance 0.08 µm CMOS," *Symp. on VLSI Technology Digest of Tech. Papers*, pp. 12-13, 1996.

[2.9] R. Gonzalez, B. Gordon, and M. Horowitz, "Supply and threshold voltage scaling for low power CMOS," *IEEE J. Solid-State Circuits*, vol. SC-32, pp. 1210-1216, August 1997.

[2.10] C. Hu, "Gate oxide scaling limits and projection," *IEEE IEDM Tech. Dig.*, pp. 319-322, 1996.

[2.11] A. Abidi, "High frequency noise measurements on FET's with small dimensions," *IEEE Trans. on Electron Devices*, vol. ED-33, pp. 1801-1805, November 1986.

[2.12] M. Pelgrom, A. Duinmaijer, and A. Welbers, "Matching properties of MOS transistors," *IEEE J. Solid-State Circuits*, vol. SC-24, pp. 1433-9, October 1989.

[2.13] D. Linden, *Handbook of Batteries*, McGraw-Hill, 1994.

[2.14] Gates Energy Products, *Rechargeable Batteries Applications Handbook*, Butterworth-Heinemann, 1992.

CHAPTER 3 Analog-to-Digital Conversion

This chapter begins with a brief overview of analog-to-digital (A/D) conversion, wherein quantization error is considered for both Nyquist-rate and oversampling A/D converters. A subclass of oversampling converters based on noise-shaping topologies commonly referred to as sigma-delta ($\Sigma\Delta$), or equivalently delta-sigma ($\Delta\Sigma$), modulators are then examined. The basic operation of a first-order $\Sigma\Delta$ modulator is described, followed by a consideration of higher-order architectures and architectures employing multibit quantization.

3.1 Analog-to-Digital Converters

The process by which an analog signal is encoded into a digital representation encompasses both the *sampling* of an analog waveform in time and *quantizing* it in amplitude. The minimum rate at which a signal can be sampled is governed by its bandwidth, while the resolution with which it is encoded dictates the amount of quantization error that can be tolerated. As depicted in Figure 3.1, in addition to the fundamental operations of sampling and quantization, the entire process of A/D conversion can also require both analog antialias filtering and digital postprocessing. Moreover, the operations of sampling and quantization need not necessarily be performed sequentially in the order shown in Figure 3.1.

Analog-to-Digital Conversion

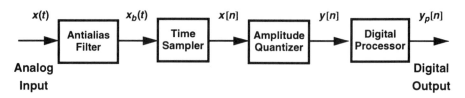

FIGURE 3.1 Block diagram of analog-to-digital conversion.

The antialias filter in Figure 3.1 is used to limit the bandwidth of the analog input signal to less than half the sampling frequency, thereby ensuring that the sampling operation will not alias noise or out-of-band signals into the baseband [3.1]. The width of the antialiasing filter's transition band increases as the converter's sampling rate is increased relative to the input bandwidth. In general, analog filters with a gentle roll off in their transition band are less costly, less complex to design, dissipate less power, are physically smaller, and introduce less phase distortion than analog filters with a steep roll off.

In the topology shown in Figure 3.1, the output of the antialias filter is sampled at uniformly spaced instances in time to produce a discrete time sequence of analog samples. The rate at which the signal is sampled relative to the signal bandwidth is a key distinguishing characteristic of A/D converters. Nyquist-rate A/D converters [3.2] derive their name from the fact that they sample the signal at approximately twice the signal bandwidth, the minimum rate that allows the reconstruction of signals as dictated by the Nyquist sampling theorem [3.1]. Such converters are well-suited to data conversion in systems wherein the conversion process is constrained by bandwidth limitations imposed by the technology in which the converter is implemented. Alternatively, oversampling A/D converters sample the input signal at a multiple of the Nyquist sampling frequency, M, referred to as the oversampling ratio. Oversampling ratios as low as eight [3.3] or as high as several hundred [3.4] have been used in such converters. As is demonstrated in Section 3.3, oversampling A/D converters exchange resolution in time for resolution in amplitude in order to ease the demands on the precision with which the signal must be quantized. They are most commonly used in applications where high resolution is needed and the signal bandwidth is much less than the bandwidth limitations imposed by the implementation technology.

The function of the digital postprocessor in Figure 3.1 depends on the architecture of the converter. It can be as simple as encoding the signal into the desired digital format, such as two's complement or Gray code, or more elaborate, such as calibration logic for a pipeline converter or the digital decimation filter in an oversampling converter.

3.1.1 Quantizer Characteristics

The performance of an A/D converter is directly influenced by the characteristics of its so-called quantizer, an element that encodes a continuous range of analog values into a set of discrete levels. A quantizer that maps its input range into 2^n output levels is referred to as an n-bit quantizer. In a *uniform* quantizer, the input signal range is partitioned into segments of equal width, each of which maps to a single output code. The input points at which the quantizer output changes are called *transition points*. The separation between the output levels is

$$\delta = \frac{\Delta}{2^n - 1}, \qquad (3.1)$$

where Δ is the maximum output range. The separation between the input levels is

$$\gamma = \frac{\Gamma}{2^n}, \qquad (3.2)$$

where Γ is the full-scale input range. The magnitude of γ is also known as the least significant bit (LSB) of the quantizer. The equivalent linear gain of a quantizer, G, is defined as

$$G = \frac{\delta}{\gamma} = \frac{\Delta(2^n)}{\Gamma(2^n - 1)}. \qquad (3.3)$$

A quantizer with an output transition at the midpoint of its input range is known as a *midrise* quantizer, while one whose output is constant at the midpoint of its input range is called a *midtread* quantizer.

Figure 3.2 shows the transfer characteristic of a 3-bit, midrise, uniform quantizer. Quantization error is defined as the difference between the quantizer output and input

$$e_Q = y - x. \qquad (3.4)$$

Analog-to-Digital Conversion

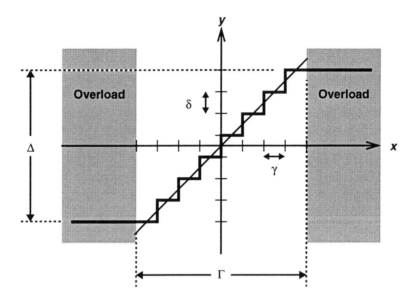

FIGURE 3.2 3-bit quantizer transfer curve.

When the input signal exceeds the full-scale input range, Γ, the quantization error grows quickly, and the quantizer is said to be in overload. The quantization error of the 3-bit quantizer of Figure 3.2 is shown in Figure 3.3.

An important characteristic of a quantizer is its resolution, also known as dynamic range, which is defined as the power[1] of a full-scale input sinusoid divided by the input-referred quantization error power. The nonlinearity in the quantizer transfer characteristic and correlation of the quantization error with the input signal complicate the exact analysis of quantizer behavior [3.5]-[3.9]. In the following paragraphs, some simplifying approximations are used to relate the dynamic range of a uniform quantizer to its resolution expressed as a number of bits.

As shown in [3.10] and [3.11], over time the spectral density of quantization error approaches that of an additive, uniformly distributed white noise that is uncorrelated with the input when the following conditions are satisfied:

---

1. In this context, the word "power" refers to the mean-square value of the signal, or, equivalently, the power dissipation normalized to a 1-Ω resistor.

Analog-to-Digital Converters

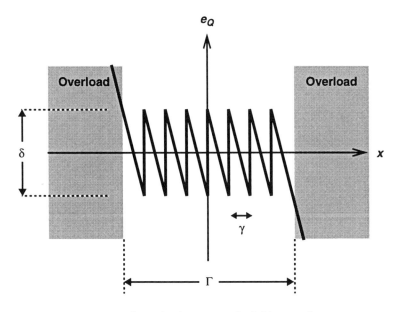

FIGURE 3.3 Quantization error of a 3-bit quantizer.

1. the input signal does not exceed the input range of the quantizer,
2. the input signal is active across many quantizer levels,
3. the joint probability density function of any two quantizer inputs is smooth, an
4. the quantizer has many quantization levels.

Herein, this noise model will be referred to as the *white noise approximation*.

A quantizer model based on the white noise approximation is shown in Figure 3.4. The gain G represents the linear gain of the quantizer, as defined in (3.3), and e_Q is an additive, uniformly distributed quantization error that is uncorrelated with the input. Under the white noise approximation, the quantizer error probability density function has the distribution shown in Figure 3.5. The power in this error is

$$\sigma_e^2 = \int_{-\infty}^{\infty} e_Q^2 \rho_E(e_Q) de_Q = \frac{\delta^2}{12}. \tag{3.5}$$

Analog-to-Digital Conversion

Analog-to-Digital Conversion

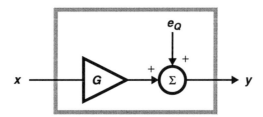

FIGURE 3.4 Quantizer model using additive noise.

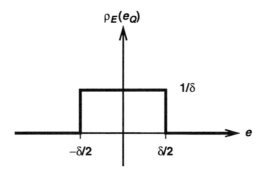

FIGURE 3.5 Quantizer error distribution under the white noise approximation.

The spectrum of the quantization noise under the white noise approximation, when the sampling rate is f_S, is shown in Figure 3.6. The power spectral density is

$$N_Q(f) = \frac{\delta^2}{12} \times \frac{1}{f_S}. \tag{3.6}$$

The power of a full-scale input sinusoid is

$$S_S = \frac{\Gamma^2}{8} = \frac{\gamma^2}{8}(2^{2n}), \tag{3.7}$$

and the input-referred quantization noise power is

Analog-to-Digital Converters

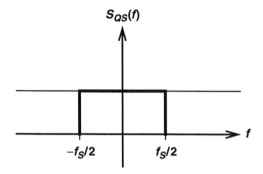

FIGURE 3.6 Quantizer power spectral density under the white noise approximation.

$$S_Q = \frac{\delta^2}{12G^2} = \frac{\gamma^2}{12}. \tag{3.8}$$

Hence, the dynamic range of an n-bit quantizer is

$$DR = \frac{S_S}{S_Q} = \frac{3}{2}(2^{2n}). \tag{3.9}$$

When expressed in dB, the dynamic range of an n-bit quantizer is

$$DR = 1.76 + 6.02n. \tag{3.10}$$

(3.9) and (3.10) indicate that a 1-bit increase in resolution corresponds to a 4-fold, or 6-dB increase in DR. For example, for a 15-bit quantizer $DR = 1.6 \times 10^9 = 92\,\text{dB}$ and for a 16-bit quantizer $DR = 6.4 \times 10^9 = 98\,\text{dB}$. The DR of an ideal, Nyquist-rate A/D converter is equal to the DR of the quantizer that it employs. As will be shown in Section 3.3, the dynamic range of an oversampling A/D converter can be much greater than that of its constituent quantizer.

Analog-to-Digital Conversion

Analog-to-Digital Conversion

3.1.2 Performance Metrics

In practice, the transfer characteristic of a quantizer differs from the ideal shown in Figure 3.2 because imperfections in a practical implementation move the transition points away from the desired levels. Nonidealities in the quantizer transfer characteristic in turn introduce errors into the A/D converter output. Performance metrics used to describe the deviation in the performance of an actual A/D converter from that of an ideal converter fall into two categories: static and dynamic. Static metrics include monotonocity, offset, gain error, differential nonlinearity (*DNL*), and integral nonlinearity (*INL*), also known as relative accuracy. A monotonic A/D converter is one for which the output never decreases when the input signal is increased. A converter's offset is the x-axis intercept of a straight line that connects the endpoints of the transfer characteristic. Gain error is the deviation in the slope of that line from the desired value. *DNL* is defined as the maximum deviation of the difference of two consecutive transition points from the ideal value, which corresponds to 1 LSB (least significant bit). A *DNL* more negative than −1 LSB means that there is at least one output code that can never be generated, which is known as a *missing code*. *INL* is defined as the maximum vertical deviation between a line connecting the endpoints of the transfer curve and the midpoint between two consecutive transition points. Both *DNL* and *INL* are typically specified in units of LSB's.

In addition to the static performance metrics that are commonly used to characterize Nyquist-rate converters, A/D converters are also characterized by dynamic metrics such as signal-to-noise ratio (*SNR*), signal-to-distortion ratio (*SDR*) or total harmonic distortion (*THD*), signal-to-noise plus distortion ratio (*SNDR*), noise floor or idle channel noise, overload level, dynamic range (*DR*), and spurious free dynamic range (*SFDR*). *SNR* is defined as the ratio of the input signal power to the power of the input-referred noise, including circuit and quantization noise but excluding the power in harmonics of the signal. In an A/D converter with uniform quantization, the *SNR* typically rises linearly with signal power until the overload level of the quantizer is reached, and then drops precipitously. *SDR* is defined as the input signal power divided by the power in the harmonics of the input signal. Since circuit nonlinearities tend to increase as signal levels increase, *SDR* typically falls as the input signal becomes large. *SNDR* is the ratio of the signal power to the sum of the noise power and the power in the harmonics of the signal. At low signal levels, *SNDR* usually follows the *SNR* curve, while at high signal levels *SNDR* follows the *SDR* curve. The noise floor is simply the input-referred baseband noise, including circuit and quantization noise. The overload level is defined as the power of an input sinusoid with amplitude larger than that which achieves the peak *SNDR*, for which the *SNDR* has fallen 3 dB below its peak value. The definition of *DR*, as stated in Section 3.1.1, may now be modified to account for implementation imperfections as follows: *DR* is equal to the overload level divided by the noise floor. *SFDR* is defined as the ratio of the signal power to the power of the strongest spectral tone, which is not necessarily harmonically related to the signal, and is measured at the input signal power that maximizes this ratio. *SFDR* is typically greater than *DR* since it does not take into account the power in broadband noise. All dynamic performance metrics are usually specified in dB.

3.2 Nyquist-Rate A/D Converters

Nyquist-rate A/D converters can be subdivided into word-at-a-time, partial-word-at-a-time, bit-at-a-time, and level-at-a-time architectures depending on the number of bits that are quantized in a single clock cycle, and thus the number of clock cycles needed to generate an output word. While there are many Nyquist-rate architectures, they all generally require that an operation such as comparison, amplification or subtraction, be performed to the overall precision of the converter. This typically translates into a need for precise component matching. In CMOS VLSI processes it is rarely practical to achieve better than 10 to 12 bits of matching [3.12], [3.13] without trimming or paying special attention to the fabrication process. Therefore, Nyquist-rate converters that deliver better than about 12 bits of resolution must rely on some form of error correction or calibration. The manufacturing and die area costs associated with such approaches to high-resolution data converter design are powerful incentives for considering alternative converter architectures that do not require precise component matching.

3.3 Oversampling A/D Converters

Oversampling A/D converters exchange resolution in time for that in amplitude by combining sampling at well above the Nyquist rate with coarse quantization embedded within a feedback loop in order to suppress the amount of quantization noise appearing in the signal band. The quantizer output is then digitally filtered to generate a lower data-rate, higher resolution encoding of the signal. The use of feedback to attenuate the noise associated with the coarse quantization is a critical aspect of oversampling A/D converters, since the improvement in dynamic range obtained without feedback is marginal. To illustrate this point, consider a Nyquist-rate converter that is

Analog-to-Digital Conversion

operated with an oversampling ratio of M. The signal bandwidth occupies only a fraction of the sampling, or Nyquist, bandwidth while, under the white noise approximation, the quantization noise is distributed uniformly across the sampling bandwidth. Therefore, by filtering the quantizer output with a brickwall lowpass digital filter so as to remove the out-of-band quantization noise and then downsampling by a factor of M, the quantization noise is reduced by a factor of M. The dynamic range thus becomes

$$DR = \frac{S_S}{S_Q/M} = \frac{3}{2}M(2^{2n}). \tag{3.11}$$

From (3.11) it is evident that an increase in resolution of one bit requires an increase in oversampling ratio by a factor of 4. Furthermore, while oversampling can reduce the baseband noise it does not improve the linearity of a converter. For example, to obtain 15 bits of resolution from an 8-bit Nyquist-rate converter, not only must the oversampling ratio be 4^7, or 16384, but the 8-bit converter must meet the linearity requirements of a 15-bit system. Due to the very high oversampling ratio needed and the severe linearity requirements, oversampling alone is not an effective means of obtaining high-resolution data conversion.

3.3.1 Feedback Modulators

The trade-off between sampling rate and resolution in an oversampling A/D converter can be made much more favorable by altering the spectrum of the quantization noise so that only a small fraction of the noise appears within the signal band. Figure 3.7 illustrates a system, referred to as a *feedback modulator*, that can be used to achieve such noise shaping. In this modulator, $A(z)$ and $F(z)$ are discrete-time filters, and quantization is accomplished with an "A/D converter" (ADC) having uniformly spaced decision levels. To establish the feedback loop, the digital output of the quantizer ADC is converted into an analog signal by means of a digital-to-analog converter (DAC). Typically, the DAC has the same resolution as the quantizer ADC and hence introduces no quantization error. Nevertheless, there may be significant implementation errors associated with the DAC. If the DAC errors can be modeled as an uncorrelated, additive white noise, the linearized model shown in Figure 3.8 may be employed to assess their impact on the performance of the modulator. The output of the system in Figure 3.8 is

$$Y(z) = \frac{GA(z)}{1 + GA(z)F(z)}X(z) + \frac{1}{1 + GA(z)F(z)}E_Q(z) - \frac{GA(z)F(z)}{1 + GA(z)F(z)}E_D(z). \tag{3.12}$$

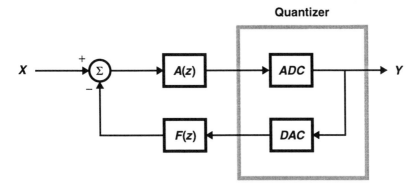

FIGURE 3.7 Block diagram of a feedback modulator.

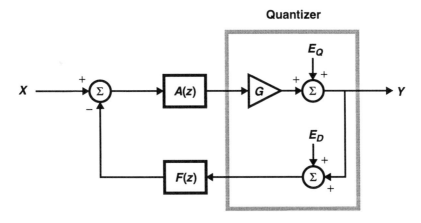

FIGURE 3.8 Linearized block diagram of a feedback modulator.

Feedback modulators are classified according to the nature of the forward and feedback transfer functions, $GA(z)$ and $F(z)$, respectively. *Delta modulators* typically use $GA(z)$ equal to unity in the baseband and an $F(z)$ with a large baseband gain in order to generate an estimate of the input signal [3.14]-[3.17]. If the estimate is close to the input, the quantizer input will be small allowing the use of a quantizer with a small input range, thereby reducing the quantization error. In a classical first-order delta modulator $F(z)$ is simply an integrator, and the modulator thus encodes the rate-of-change in the input signal, rather than the signal itself. In *Interpolative modulators*, both the forward path and feedback transfer functions have a large baseband gain [3.18]-[3.20]. This approach allows the use of a quantizer with a small input range and, at the same time, attenuates the baseband quantization noise. In both of these approaches, the large baseband gain in the feedback transfer function amplifies baseband DAC errors relative to the input signal. Thus, the subsequent digital coding operation used to recover the signal has the effect of greatly amplifying E_D, which presents serious implementation difficulties.

Sigma-delta ($\Sigma\Delta$) *modulators* are a subclass of feedback modulators that avoid amplifying DAC errors by setting the baseband feedback gain equal to unity and relying on a large baseband gain in the forward path to attenuate the quantization error [3.21], [3.22]. However, even in $\Sigma\Delta$ modulators, DAC errors remain an important concern since they add directly to the modulator input and are, therefore, indistinguishable from the signal. In particular, the linearity of the DAC must meet the overall linearity specification of the modulator, which is a severe requirement in high-resolution applications.

In order to overcome the stringent linearity requirements for the DAC, a 1-bit quantizer is frequently used in $\Sigma\Delta$ modulators. Since a 1-bit quantizer has only two output levels, deviations in these levels from the desired values simply introduce offset and gain errors rather than nonlinearity [3.4]. For this reason, most of the attention in this text, with the exception of Section 3.3.6, is devoted to modulators that employ 1-bit quantizers. The use of a 1-bit quantizer, while advantageous from a DAC linearity perspective, does entail several important ramifications. Most obviously, the error introduced by a 1-bit quantizer is quite large. In the following section it will be shown that $\Sigma\Delta$ modulators modify the spectral shape of this error so as to ensure that the baseband noise power is small. Thus, $\Sigma\Delta$ modulators are commonly referred to as noise-shaping modulators.

An important consequence of using a 1-bit quantizer is that the assumptions upon which the white noise approximation is based are violated. The breakdown of the white noise approximation is evident in the fact that an analysis based on this approximation fails to predict spectral tones that are inherent in many 1-bit $\Sigma\Delta$ modulators [3.4]. These tones are a consequence of repeating patterns in the output bit stream. They can appear in response to dc, as well as ac, inputs and can occur at frequencies within the signal bandwidth. This issue is addressed in Chapter 5 and plays an important role in the selection of a modulator architecture. Nevertheless, simulations and empirical evidence have shown that the white noise approximation remains a valuable

Oversampling A/D Converters

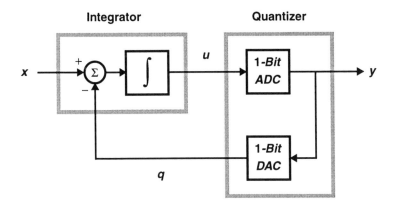

FIGURE 3.9 A first-order, 1-bit sigma-delta modulator.

tool in designing $\Sigma\Delta$ modulators when its use is coupled with extensive computer simulations.

3.3.2 First-Order Sigma-Delta Modulators

Shown in Figure 3.9 is a simple, first-order, 1-bit $\Sigma\Delta$ modulator. It comprises a discrete-time difference integrator and a 1-bit quantizer that maps its input to both an analog and a digital output. The digital output of the modulator is

$$y[n] = Q\{u[n]\}, \tag{3.13}$$

where

$$u[n] = u[n-1] + x[n-1] - Q_a\{u[n-1]\}, \tag{3.14}$$

$$q[n] = Q_a\{u[n]\}, \tag{3.15}$$

$$Q\{u[n]\} = \begin{cases} 1, & u[n] \geq 0 \\ 0, & u[n] < 0, \end{cases} \tag{3.16}$$

Analog-to-Digital Conversion   **33**

Analog-to-Digital Conversion

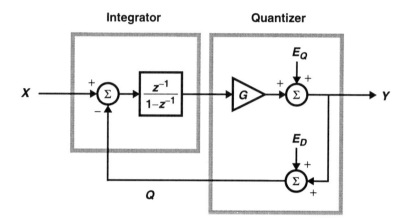

FIGURE 3.10 Linearized model of a first-order sigma-delta modulator.

$$Q_a\{u[n]\} = \begin{cases} \dfrac{\Gamma}{2}, & u[n] \geq 0 \\ -\dfrac{\Gamma}{2}, & u[n] < 0 \end{cases} \quad (3.17)$$

$Q\{\cdot\}$ represents analog-to-digital quantization, $Q_a\{\cdot\}$ represents analog-to-analog quantization, and Γ is the output range of the 1-bit DAC.

The discrete-time difference integrator accumulates the difference between the input signal, $x[n]$, and the DAC output level, $q[n]$, at its output, $u[n]$. When the integrator output, $u[n]$, crosses the quantizer threshold, the sign of $q[n]$ changes, reversing the polarity of the integrator input if $|x[n]| < |q[n]|$. This causes the integrator output to move in the opposite direction in the following cycle. Thus, the negative feedback loop tries to force $q[n]$ to equal $x[n]$. However, the coarseness of the quantizer output causes $q[n]$ to oscillate so that only its time average approaches $x[n]$. Therefore, the modulator output, $y[n]$, is a digital bit stream whose average value is a digital approximation of the input signal. Note that if $|x[n]| > |q[n]|$ the feedback loop is unable to change the polarity of the integrator input and the modulator output can no longer track the input signal. This condition is known as *modulator overload*.

By applying the white noise approximation, the modulator of Figure 3.9 may be represented by the linear system shown in Figure 3.10. Unfortunately, the definition of quantizer gain, G, in (3.3) breaks down for a 1-bit quantizer because the input step size, γ, is undefined. Physically, this quandary is illustrated by the fact that the 1-bit quantizer is insensitive to any gain that precedes it, since it only detects the polarity of its input. An approximation for the quantizer gain, which will be referred to herein as the *unity gain approximation*, is to assume a gain, G, such that the product of the integrator gain (in this case unity) and the quantizer gain around the outermost feedback loop of a modulator is equal to 1. It should be stressed that the only justification for this approximation is that analytical results obtained under this approximation are in good agreement with computer simulations, analytical modeling [3.23], and experimental measurements.

In the first-order $\Sigma\Delta$ modulator of Figure 3.10, if the unity gain approximation is adopted, the quantizer gain is 1. With this definition, the z-domain output can be derived as

$$Y(z) = z^{-1}X(z) + (1-z^{-1})E_Q(z) - z^{-1}E_D(z). \tag{3.18}$$

It is apparent from (3.18) that the modulator output, $Y(z)$, has three components. One is the input signal delayed by one sample. Another is the quantizer error filtered by a first-order difference, which has the transfer function $H_Q(z) = 1-z^{-1}$. This filtering, referred to as *noise shaping*, places a zero at dc, thereby attenuating the low-frequency quantization noise. When the shaping of the quantization noise is accomplished by a differencing operation, this specific class of modulators is referred to as *noise-differencing* modulators. The third component of the output is the DAC error delayed by one sample but otherwise unaltered. As mentioned earlier, the use of a 1-bit DAC ensures that this is a linear error.

To derive an expression for the dynamic range of a first-order $\Sigma\Delta$ modulator, the power of a full-scale input signal is first calculated. Next, the power of the quantization noise within the signal baseband is determined by making use of the white noise approximation. In these calculations, it is assumed that the quantization noise lying outside the baseband is eliminated by a lowpass digital filter following the modulator and thus does not appear in the baseband after downsampling.

For the purposes of this text, a full-scale input to a $\Sigma\Delta$ modulator is defined as one with a peak-to-peak amplitude equal to the output range of the quantizer DAC, in this case Γ. According to this definition, a full-scale input sinusoid has a power of

Analog-to-Digital Conversion

$$S_S = \frac{\Gamma^2}{8}. \tag{3.19}$$

The magnitude of the noise shaping transfer function is

$$\begin{aligned}|H_Q(f)| &= \sqrt{|H_Q(z)|^2_{z=e^{j2\pi fT}}} \\ &= \sqrt{(1-e^{-j2\pi fT})(1-e^{j2\pi fT})} \\ &= \sqrt{2-2\cos(2\pi fT)} \\ &= 2\sin(\pi fT).\end{aligned} \tag{3.20}$$

It follows from (3.6) and (3.20) that the power spectral density of the input-referred quantization noise in a first-order $\Sigma\Delta$ modulator is

$$N_{QS}(f) = (2\sin(\pi fT))^2 \times \frac{\Gamma^2}{12 f_S}. \tag{3.21}$$

The power of the baseband quantization noise is obtained by integrating (3.21) over the Nyquist signal bandwidth

$$S_Q = \int_{-f_B}^{f_B} N_{QS}(f) df \approx \frac{\pi^2}{3} \frac{1}{M^3} \frac{\Gamma^2}{12}, \tag{3.22}$$

where $f_B = f_N/2$ is the signal bandwidth, $M = f_S/f_N$ is the oversampling ratio, and $\sin(\pi fT)$ is approximated as πfT, which is valid for $M \gg 1$. The dynamic range of the first-order $\Sigma\Delta$ modulator can be found by dividing (3.19) by (3.22) to obtain

$$DR = \frac{S_S}{S_Q} = \frac{9M^3}{2\pi^2}. \tag{3.23}$$

Thus, the dynamic range increases by 9 dB, or 1.5 bits, per octave increase in M.

Simulations show that the operation of the first-order $\Sigma\Delta$ modulator architecture is highly tolerant of imperfections in the constituent components such as quantizer offset and integrator gain error. However, an important consideration in oversampling modulators employing 1-bit quantization that is not predicted by the white noise approximation and linear analysis is the presence of spurious tones in the output spectrum. These tones are particularly strong in a first-order modulator [3.6]-[3.9]. Another limitation of the first-order architecture is that, while the trade-off between M and DR as expressed in (3.23) is a significant improvement over (3.11), high-resolution data conversion still requires a relatively high oversampling ratio. For example, to achieve a resolution of 16 bits with the modulator of Figure 3.9, M must be at least 2400. Both of these limitations are much less severe in higher-order $\Sigma\Delta$ modulators.

3.3.3 Higher-Order Noise-Differencing Sigma-Delta Modulators

Transfer functions for high-order, noise-differencing modulators may be derived by considering the signal and quantization noise transfer functions presented in (3.12) for Figure 3.8. If the DAC error is neglected, (3.12) may be rewritten as

$$Y(z) = H_X(z)X(z) + H_Q(z)E_Q(z), \qquad (3.24)$$

where

$$H_X(z) = \frac{GA(z)}{1 + GA(z)F(z)}, \qquad (3.25)$$

and

$$H_Q(z) = \frac{1}{1 + GA(z)F(z)}. \qquad (3.26)$$

It follows that

$$GA(z) = \frac{H_X(z)}{H_Q(z)}, \qquad (3.27)$$

and

$$F(z) = \frac{1 - H_Q(z)}{H_X(z)}. \qquad (3.28)$$

In an Lth-order noise-differencing modulator,

Analog-to-Digital Conversion

$$H_Q(z) = (1 - z^{-1})^L, \qquad (3.29)$$

and

$$H_X(z) = z^{-L}. \qquad (3.30)$$

Thus, the quantization noise is attenuated by L zeros at dc, and the input signal is delayed but spectrally unaltered. The exponent of z in (3.30) may be any integer between $-L$ and -1; a value of $-L$ simply makes an implementation using switched-capacitor circuits especially straightforward. To realize a second-order noise-differencing modulator, (3.27) and (3.28) are evaluated when $L = 2$ to yield

$$GA(z) = \frac{z^{-2}}{(1 - z^{-1})^2}, \qquad (3.31)$$

and

$$F(z) = \frac{2 - z^{-1}}{z^{-1}} = 2z - 1. \qquad (3.32)$$

Figure 3.11(a) illustrates the block diagram of a linearized, discrete-time, second-order, noise-differencing, $\Sigma\Delta$ modulator that uses (3.31) and (3.32) for the forward and feedback transfer functions, respectively. This topology can be simplified considerably by manipulating its various elements while preserving its transfer functions. The manipulations proceed as follows. First, as shown in Figure 3.11(b), the feedback network is rearranged so that there is a unity gain signal path along the feedback loop. Next, the component of the feedback transfer function not in the unity gain loop is connected to an intermediate node in the forward path transfer function immediately following the first integrator, as shown in Figure 3.11(c). To preserve the system transfer function, an integrator must be added to the nested feedback loop. Since the differentiation and integration in the nested loop cancel one another, the system can be simplified to that shown in Figure 3.12. The unity gain approximation states that the product of the integrator gains along the outer feedback loop and the effective quantizer gain, G, is unity. Thus, under the unity gain approximation, G in Figure 3.12 is equal to 1, and (3.31) is satisfied.

The dynamic range of the second-order architecture may be calculated in a manner similar to that used for the first-order modulator. It can easily be shown that

Oversampling A/D Converters

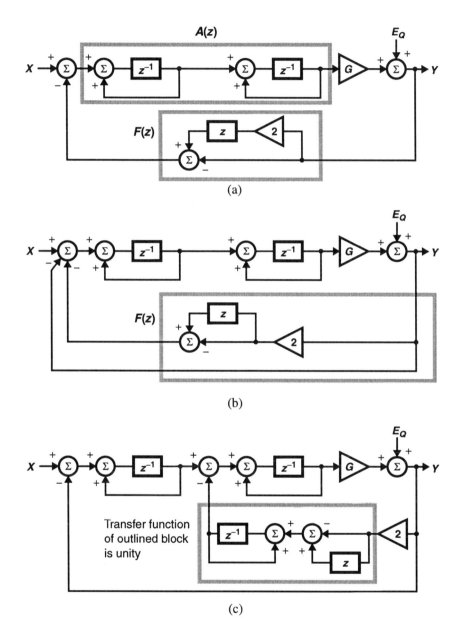

FIGURE 3.11 Block diagram manipulation of the second-order sigma-delta modulator.

Analog-to-Digital Conversion

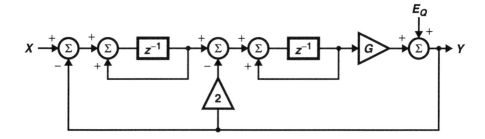

FIGURE 3.12 Simplified block diagram of the second-order sigma-delta modulator.

$$DR = \frac{S_S}{S_Q} = \frac{15M^5}{2\pi^4}. \qquad (3.33)$$

Thus, the dynamic range increases by 15 dB, or 2.5 bits, per octave increase in M. The increased noise shaping is a key advantage of the second-order modulator over the first-order modulator. For example, a dynamic range of 98 dB, or 16 bits, may be obtained with an oversampling ratio of only 152. Another significant benefit of the second-order architecture is a large reduction in the number and amplitude of spurious quantization noise tones observed in simulations. In effect, the increased filtering in the forward path of the modulator serves to reduce the occurrence of repeating patterns in the output bit stream. Furthermore, behavioral simulations reveal that the second-order architecture is nearly as tolerant of imperfections in its constituent components as the first-order topology [3.24].

For the modulator of Figure 3.12, simulations show that the output signal range of the integrators is several times larger than the full-scale modulator input range. This requirement presents a significant challenge in the implementation of the modulator in a technology where the dynamic range is constrained. To relax the integrator signal range requirements, integrator gains that reduce signal swings may be employed. Shown in Figure 3.13 is the block diagram of a second-order ΣΔ modulator that has been generalized to include independent gains for each of the integrator inputs. In a switched-capacitor circuit implementation these gains are easily realized by appropriately sizing the input sampling capacitors.

According to the unity gain approximation, the equivalent quantizer gain in the topology of Figure 3.13 is

Oversampling A/D Converters

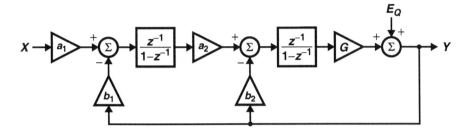

FIGURE 3.13 Second-order sigma-delta modulator with integrator signal scaling.

$$G = \frac{1}{b_1 a_2}. \tag{3.34}$$

To implement the same noise-differencing transfer function provided by the modulator in Figure 3.12, the integrator gains in the modulator of Figure 3.13 must satisfy the following constraint:

$$\frac{b_2}{b_1 a_2} = 2. \tag{3.35}$$

In addition, this modulator scales the input signal relative to the quantizer output range by

$$G_{A/D} = \frac{a_1}{b_1}. \tag{3.36}$$

Since $G_{A/D}$ is chosen on the basis of overall system considerations, the modulator of Figure 3.13 adds two degrees of freedom that may be used to tailor the output swing of the integrators. In particular, the modulator may be designed so that the output range required for the first and second integrators is approximately the same as the analog output range of the quantizer. The second-order, noise differencing architecture has been used successfully in a variety of applications ranging from voiceband telephony to digital-audio systems [3.25], [3.26].

Modulators of order greater than 2 provide more attenuation of the baseband quantization noise than the second-order architecture and thus allow the use of lower oversampling ratios. This increases the signal bandwidth that can be accommodated for a

Analog-to-Digital Conversion

Analog-to-Digital Conversion

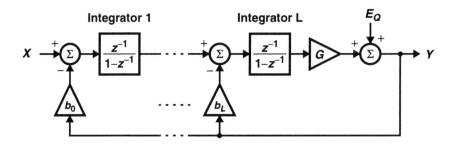

FIGURE 3.14 Block diagram of an Lth-order sigma-delta modulator.

given sampling rate, and can make $\Sigma\Delta$ modulation a feasible approach for higher bandwidth applications. In some cases, higher-order modulators also reduce unwanted spectral tones to levels well below those that occur in the second-order architecture. One approach to defining a higher-order, noise-differencing $\Sigma\Delta$ modulators is simply to proceed in the same fashion used to derive the second-order topology. To implement a noise-differencing modulator, the transfer functions in the system of Figure 3.8 are

$$GA(z) = \frac{z^{-L}}{(1-z^{-1})^L} \qquad (3.37)$$

and

$$F(z) = \frac{1-(1-z^{-1})^L}{z^{-L}}. \qquad (3.38)$$

It can be shown that the modulator of Figure 3.14 implements an Lth-order noise differencing $\Sigma\Delta$ modulator if and only if the terms b_n are the binomial coefficients [3.23].

The dynamic range of the Lth-order architecture may be estimated in a manner similar to that used for the first- and second-order architectures. It follows that

$$DR = \frac{3}{2}\left(\frac{2L+1}{\pi^{2L}}\right)M^{2L+1}. \qquad (3.39)$$

Thus, in an Lth-order modulator, *DR* increases by $3+6L$ dB, or $0.5+L$ bits, per octave increase in *M*.

Unfortunately, unlike first- and second-order modulators, 1-bit noise-differencing ΣΔ modulators of third and higher order are inherently unstable [3.4]. This instability manifests itself as unbounded growth of integrator output levels with the input bounded by some predetermined fraction of the feedback reference voltage range. To achieve higher than second-order noise shaping in a 1-bit, single-stage design, noise transfer functions other than differencing must be used.

3.3.4 Higher-Order Single-Stage Sigma-Delta Modulators

Many approaches have been devised to implement stable, higher-order, single-stage, 1-bit ΣΔ modulation [3.27]-[3.30]. The general idea in these architectures is to limit the magnitude of the noise transfer function at high frequencies to a level that stabilizes the system, as determined by simulations. While this approach can be effective in stabilizing a modulator, it also typically increases the inband noise relative to that reflected in (3.39). A characteristic that is shared among these techniques is that the maximum input signal level must be restricted to a fraction of the feedback voltage range. For example, in the fourth-order modulator of [3.27] with an oversampling ratio of 80, *DR* = 104 dB, the peak *SNDR* is 96 dB, and the maximum allowed input power is −4 dB with respect to an input signal with a peak-to-peak amplitude equal to the difference between the two quantizer feedback levels. In the third-order modulator of [3.29], with an oversampling ratio of 80, *DR* = 103 dB, the peak *SNDR* is 101 dB, and the full-scale input power is −8 dB relative to difference between the quantizer output levels. In addition to constraining the input signal range, approaches used to stabilize higher-order, single-loop, 1-bit modulators require the use of either circuits that detect when overload occurs and then reset the integrators or limiters that restrict the integrator output range. Finally, simulations have shown that some higher-order, single-stage, 1-bit modulators are not able to fully eliminate spectral tones that arise from the correlation of the quantization noise with the input [3.31]. Alternatively, cascaded architectures can be use to avoid stability concerns, provide dynamic range close to that given in (3.39) for an ideal noise-differencing modulator, and suppress spurious inband tones [3.31].

3.3.5 Cascaded Sigma-Delta Modulators

High-order ΣΔ modulators may be implemented using a cascade of low-order stages. The input of the succeeding stages is driven by a scaled, and perhaps asymmetrically

Analog-to-Digital Conversion

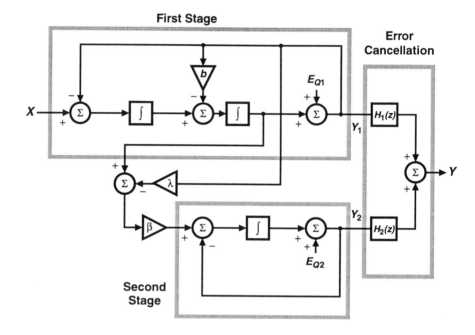

FIGURE 3.15 Block diagram of a 2-1 cascaded sigma-delta modulator.

weighted, representation of the quantization error of the preceding stage. The output of each stage can then be used to cancel the quantization error of the preceding stage. Thus, in an ideal cascade the only quantization noise appearing in the output is that of the last stage, attenuated by a noise shaping function of order equal to the overall order of the cascade [3.32]-[3.35]. The architectures of cascaded $\Sigma\Delta$ modulators are typically referred to by a sequence of numbers that represent the order of noise shaping provided by each stage in the cascade [3.35]. For example, the third-order architecture of Figure 3.15 is referred to as a 2-1 architecture because it comprises a second-order stage followed by a first-order stage. Cascaded third-order $\Sigma\Delta$ modulation may also be accomplished with 1-1-1 and 1-2 architectures. However, it can shown that the most robust circuit implementations are obtained when the first stage of the cascade is a second-order modulator [3.35].

Figure 3.15 is a linearized block diagram for a 2-1 cascaded modulator. Under the white noise and unity gain approximations, the z-domain output of the first stage of this modulator when $b = 2$ is

$$Y_1(z) = z^{-2}X(z) + (1-z^{-1})^2 E_{Q1}(z), \tag{3.40}$$

and the output of the second stage, $Y_2(z)$, is

$$Y_2(z) = z^{-1}X_2(z) + (1-z^{-1})E_{Q2}(z), \tag{3.41}$$

where

$$X_2(z) = \beta[(1-\lambda)Y_1(z) - E_{Q1}(z)]. \tag{3.42}$$

The factor λ is referred to as the error mixing coefficient, and β is known as the error gain coefficient. It is shown in Chapter 5 that these coefficients can be used to adjust the overload characteristics and the quantization noise floor of the cascaded modulator.

The overall output of the modulator in Figure 3.15 is

$$\begin{aligned}Y(z) =\ & [z^{-2}H_1(z) + z^{-3}H_2(z)\beta(1-\lambda)]X(z) \\ & + [(1-z^{-1})^2(H_1(z) + z^{-1}H_2(z)\beta(1-\lambda)) - z^{-1}H_2(z)\beta]E_{Q1}(z) \\ & + (1-z^{-1})H_2(z)E_{Q2}(z).\end{aligned} \tag{3.43}$$

The quantizer error of the first-stage, $E_{Q1}(z)$, is cancelled if the transfer functions of the error cancellation filters are

$$H_1(z) = z^{-1} - (1-\hat{\lambda})(1-z^{-1})^2 z^{-1}, \tag{3.44}$$

$$H_2(z) = \frac{1}{\hat{\beta}}(1-z^{-1})^2, \tag{3.45}$$

where $\hat{\lambda}$ and $\hat{\beta}$ are digital estimates of the analog coefficients, λ and β, respectively. Variations in the circuit component values that determine these parameters result in small differences between the analog coefficients and their digital counterparts. These matching errors, denoted herein by δ_β and δ_λ, are defined such that

$$\beta = \hat{\beta}(1+\delta_\beta), \tag{3.46}$$

Analog-to-Digital Conversion

$$\lambda = \hat{\lambda}(1 + \delta_\lambda). \tag{3.47}$$

If higher-order difference terms are neglected, it follows from (3.43)-(3.47) that the modulator output is

$$Y(z) = z^{-3}X(z) - \delta_\beta(1-z^{-1})^2 z^{-1} E_{Q1}(z) + \frac{1}{\hat{\beta}}(1-z^{-1})^3 E_{Q2}(z). \tag{3.48}$$

(3.48) indicates that, in the absence of matching errors, quantization noise in the output of the 2-1 cascade consists only of the noise introduced by the second-stage quantizer, $E_{Q2}(z)$, shaped by third-order noise-differencing and divided by $\hat{\beta}$. Since typical values for $\hat{\beta}$ are 0.5 to 0.25, this system provides nearly ideal third-order noise differencing performance. Furthermore, since it accomplishes this without the use of a high-order single-quantizer loop, the full-scale input signal amplitude need not be constrained in order to insure stability. While the modulator is relatively insensitive to matching errors in estimating λ, errors in estimating β cause noise from the first stage quantizer, $E_{Q1}(z)$, to leak to the output with second-order shaping.

The tolerance the 2-1 cascade to mismatch errors can be quantified by considering the relative contribution of the first- and second-stage quantizers to the overall noise in the modulator output. It follows from (3.48) that the total baseband quantization noise power is

$$S_{ee} = \delta_\beta^2 \frac{\pi^4}{5M^5}\sigma_{Q1}^2 + \frac{1}{\hat{\beta}^2}\frac{\pi^6}{7M^7}\sigma_{Q2}^2, \tag{3.49}$$

where σ_{Q1}^2 and σ_{Q2}^2 are the quantization noise powers of the first- and second-stage quantizers, respectively. From (3.49), it follows that in order to limit the increase in baseband noise to less than 1 dB above that obtained with perfect matching, the following condition must be satisfied

$$\begin{aligned}\delta_\beta &< \sqrt{\frac{5(10^{0.1}-1)}{7}}\frac{\pi}{\hat{\beta}M} \\ &< 0.43\frac{\pi}{\hat{\beta}M}.\end{aligned} \tag{3.50}$$

For example, with $\beta = 0.25$ and $M = 80$, the matching error, δ_β, must be less than 6.8%. This level of matching is easily realizable in modern CMOS VLSI technologies

[3.12]. In general, as the dynamic range required of the overall modulator is increased beyond that provided by the first stage, the matching constraints become more severe. In an Lth-order cascaded modulator employing a second-order section as the first stage, the matching required is inversely proportional to M^{L-2}.

Another important nonideality that can degrade the performance of a cascaded $\Sigma\Delta$ modulator is integrator leak. Integrator leak is a mechanism by which integrator state information is lost over time. The leak is a consequence of finite integrator gain at dc and can limit the extent to which a modulator can shape its quantization noise. The transfer function of a delaying, leaky integrator is

$$H(z) = \frac{z^{-1}}{1 - (1 - \varepsilon)z^{-1}}, \quad (3.51)$$

where the leakage factor, ε, is much less than one. When the modulator of Figure 3.15 is implemented with leaky integrators, the modulator output, $Y(z)$, is approximately

$$Y(z) = z^{-3}X(z)$$
$$+ z^{-2}[(1-z^{-1})(\varepsilon_1 + \varepsilon_2) + z^{-1}\varepsilon_1\varepsilon_2]E_{Q1}(z) \quad (3.52)$$
$$+ \frac{1}{\hat{\beta}}[(1-z^{-1})^3 + (1-z^{-1})^2 z^{-1}\varepsilon_3]E_{Q2}(z),$$

where ε_1, ε_2, and ε_3 are the leakage factors of the first, second, and third integrators, respectively. It follows from (3.52) that the total baseband quantization noise power is

$$S_{ee} = \left[\frac{\varepsilon_1^2 \varepsilon_2^2}{M} + (\varepsilon_1 + \varepsilon_2)^2 \frac{\pi^2}{3M^3}\right]\sigma_{Q1}^2 + \frac{1}{\hat{\beta}^2}\left[\varepsilon_3^2 \frac{\pi^4}{5M^5} + \frac{\pi^6}{7M^7}\right]\sigma_{Q2}^2, \quad (3.53)$$

where σ_{Q1}^2 and σ_{Q2}^2 are the quantization noise powers of the first- and second-stage quantizers, respectively. If the following condition holds

$$M \ll \frac{\pi}{\sqrt{3}}\left(\frac{1}{\varepsilon_1} + \frac{1}{\varepsilon_2}\right), \quad (3.54)$$

then (3.53) reduces to

Analog-to-Digital Conversion

$$S_{ee} = (\varepsilon_1 + \varepsilon_2)^2 \frac{\pi^2}{3M^3}\sigma_{Q1}^2 + \frac{1}{\hat{\beta}^2}\left[\varepsilon_3^2 \frac{\pi^4}{5M^5} + \frac{\pi^6}{7M^7}\right]\sigma_{Q2}^2. \qquad (3.55)$$

The tolerance of the 2-1 cascade to integrator leak can be quantified by considering the relative contribution of integrator leakage to the overall noise in the modulator output. It follows from (3.55) that in order to limit the increase in baseband quantization noise due to the leak of each integrator to less than 1 dB beyond that obtained with ideal integrators, the following conditions must be satisfied

$$\varepsilon_1, \varepsilon_2 < \sqrt{\frac{3(10^{0.1} - 1)}{7}} \frac{\pi^2}{\hat{\beta} M^2}$$

$$< 0.33 \frac{\pi^2}{\hat{\beta} M^2}, \qquad (3.56)$$

$$\varepsilon_3 < \sqrt{\frac{5(10^{0.1} - 1)}{7}} \frac{\pi}{M}$$

$$< 0.43 \frac{\pi}{M}. \qquad (3.57)$$

With $\hat{\beta} = 0.25$ and $M = 80$, ε_1 and ε_2 must be less than 0.002, and ε_3 must be less than 0.017. In a switched-capacitor integrator with a gain of unity, these leakage terms correspond to a requirement for operational amplifier gains of 500 and 59 in the first and second stages, respectively. Open loop gains of this magnitude can readily be achieved in a standard CMOS VLSI technology [3.36].

The dynamic range of the 2-1 modulator under the white noise approximation, when matching errors and integrator leakage do not limit the performance, is

$$DR = \frac{21\beta^2 M^7}{2\pi^6}. \qquad (3.58)$$

Thus, the dynamic range increases by 21 dB, or 3.5 bits, per octave increase in oversampling ratio, M. When $\beta = 0.25$, 16-bit resolution can be attained with $M = 71$. A particularly advantageous characteristic of the cascade architecture is that the modulator output is nearly free of discrete tones. This architecture is considered in greater detail in Section 5.1.1, with particular attention given to the choice of b, β, and λ and to the selection of the integrator gains.

3.3.6 Multibit Sigma-Delta Modulators

The discussions in this section have to this point focused exclusively on $\Sigma\Delta$ modulators that employ 1-bit quantization. As mentioned in Section 3.3.1, the use of a 1-bit quantizer greatly simplifies the task of achieving linearity in the feedback DAC equal to that required of the overall modulator since a 1-bit DAC is inherently linear. However, this benefit is accompanied by a number of penalties. The quantization noise introduced by a 1-bit quantizer is necessarily quite large and necessitates the use of a relatively high oversampling ratio in order to adequately suppress the noise power appearing in the baseband. The use of a 1-bit quantizer is also responsible for the stability problems in single-stage modulators of order greater than 2. Analytical studies have shown that higher-order, single-stage, noise-differencing modulators are stable if the number of bits in the quantizer is equal to or greater than the order of the modulator [3.37]. Moreover, spurious noise tones can be significantly reduced by the use of a multibit quantizer.

The dynamic range of an Lth-order, multibit $\Sigma\Delta$ modulator can be estimated using the white noise approximation. Under this approximation, increasing the resolution of the quantizer simply decreases the power of the quantization noise. It follows that, in an Lth-order multibit noise-differencing $\Sigma\Delta$ modulator,

$$DR = \frac{3}{2}\left(\frac{2L+1}{\pi^{2L}}\right)M^{2L+1}(2^B - 1)^2, \qquad (3.59)$$

where B is the number of bits in the quantizer. A key attribute of multibit quantizers is that the enhancement of the modulator resolution is independent of the oversampling ratio. Therefore, this technique is particularly useful in high-speed applications where a high oversampling ratio cannot be used [3.38].

The benefits of multibit quantizers have stimulated interest in techniques that can reduce the impact of nonlinearity in the quantizer DAC on the performance of a $\Sigma\Delta$ modulator. Proposed approaches include analog calibration [3.39], digital correction [3.40]-[3.42], noise shaping in a cascaded architecture [3.38], and data-dependent DAC element selection [3.43]-[3.52]. Each of these techniques has its limitations. For example, analog calibration and digital correction result in increased circuit complexity and require a test cycle to characterize the DAC. Performance increases obtained through the use of a multibit DAC in a cascaded modulator come at the expense of more stringent matching and integrator leak requirements. Finally, data-dependent DAC element selection algorithms generally suffer from both complexity and the occurrence of unwanted spectral tones.

3.4 Summary

A/D converters can be classified according to the rate at which they sample their analog input and the means by which they attain a specified resolution. Nyquist-rate A/D converters sample the input signal at close to the minimum sampling rate that can used without aliasing out-of-band signals and noise into the signal band. The resolution of a Nyquist-rate A/D converter can only be increased by quantizing the input signal in finer increments. In an oversampling A/D converter the input is sampled at a rate well above the Nyquist rate, and then filtering and negative feedback are used to shape the noise introduced by a coarse quantizer so that most of its energy lies outside the signal band. A digital lowpass decimation filter that removes the out-of-band noise allows resampling of the output at the Nyquist rate.

Oversampling A/D converters can be classified according to the transfer functions they implement for the signal and the quantization noise. A/D converters based on $\Sigma\Delta$ modulation are a robust family of oversampling converters that pass the input signal essentially unaltered, but suppress the quantization noise within the signal baseband. As the order of noise filtering is increased, the oversampling ratio need to achieve specified resolution is reduced. Unfortunately, the design of modulators of order greater than 2 is complicated by stability considerations. Stability problems encountered in higher-order, single-stage converters may be avoided by using a cascaded architecture. Cascaded modulators generally impose more stringent matching and integrator leak requirements on their implementations than single-stage architectures. However, for most applications these requirements can easily be met using standard integrated circuit technologies. Another means of extending the resolution of a $\Sigma\Delta$ modulator is to use a multibit quantizer. However, a significant drawback of this approach is that the feedback DAC must meet the full linearity requirement of the converter. Techniques for linearizing the DAC are an active research topic.

REFERENCES

[3.1] A. Oppenheim and R. Schafer, *Discrete-Time Signal Processing*, Prentice-Hall, 1989.

[3.2] B. Razavi, *Principles of Data Conversion System Design*, IEEE Press, 1995.

[3.3] T. Brooks, et al., "A 16b $\Sigma\Delta$ pipeline ADC with 2.5MHz output data-rate," *ISSCC Digest of Tech. Papers*, pp. 208-209, February 1997.

[3.4] J. Candy and G. Temes, "Oversampling methods for A/D and D/A conversion," in *Oversampling Delta-Sigma Data Converters*, pp. 1-29, New York: IEEE Press, 1992.

[3.5] N. Blachman, "The intermodulation and distortion due to quantization of sinusoids," *IEEE Trans. on Acoustics, Speech, and Signal Processing*, vol. ASSP-33, pp. 1417-1426, December 1985.

[3.6] R. Gray, "Quantization noise spectra," *IEEE Trans. Information Theory*, vol. 36, pp. 1220–1244, November 1990.

[3.7] R. Gray, "Spectral analysis of quantization noise in a single-loop sigma-delta modulator with dc input," *IEEE Trans. on Communications*, vol. 37, pp. 588–599, June 1989.

[3.8] W. Chou, P. Wong, and R. Gray, "Multistage sigma-delta modulation," *IEEE Trans. on Information Theory*, vol. 35, pp. 784–796, July 1989.

[3.9] R. Gray, W. Chou, and P. Wong, "Quantization noise in single-loop sigma-delta modulation with sinusoidal inputs," *IEEE Trans. on Communications*, vol. 37, pp. 956–968, September 1989.

[3.10] W. Bennett, "Spectra of quantized signals," *Bell System Tech. Journal*, vol. 27, pp. 446–472, 1948.

[3.11] B. Widrow, "A study of rough amplitude quantization by means of Nyquist sampling theory," *IRE Trans. Circuit Theory*, vol. CT-3, pp. 266–276, December 1956.

[3.12] M. Pelgrom, A. Duinmaijer, and A. Welbers, "Matching properties of MOS transistors," *IEEE J. Solid-State Circuits*, vol. SC-24, pp. 1433-9, October 1989.

[3.13] T. Mizuno, J. Okamura, and A. Toriumi, "Experimental study of threshold voltage fluctuation due to statistical variation of channel dopant number in MOSFET's," *IEEE Trans. on Electron Devices*, vol. ED-41, pp. 2216-2221, November 1994.

[3.14] F. de Jager, "Delta modulation, a method of PCM transmission using the 1-unit code," *Philips Res. Rep.*, vol. 7, pp. 442-466, 1952.

[3.15] A. Tomozawa and H. Kaneko, "Companded delta modulation for telephone transmission," *IEEE Trans. Comm. Tech.*, vol. COM-16, February 1968.

[3.16] J. Greefkes and F. de Jager, "Continuous delta modulation," *Philips Res. Rep.*, vol. 23, pp. 442-466, 1952.

[3.17] N. Jayant, "Adaptive delta modulation with a one bit memory," *Bell Syst. Tech. J.*, March 1970.

[3.18] J. Candy, W. Ninke, and B. Wooley, "A per-channel A/D converter having 15-segment µ-255 companding," *IEEE Trans. Commun.*, vol. COM-24, pp. 33-42, January 1976.

[3.19] B. Wooley and J. Henry, "An integrated per-channel PCM encoder based on interpolation," *IEEE J. Solid-State Circuits*, vol. SC-14, pp. 14-20, February 1979.

[3.20] J. Candy, B. Wooley, and O. Benjamin, "A voiceband codec with digital filtering," *IEEE Trans. Commun.*, vol. COM-29, pp. 815-830, June 1981.

[3.21] C. Cutler, "Transmission systems employing quantization," 1960 US. Patent No. 2,927,962 (filed 1954).

[3.22] H. Inose and Y. Yasuda, "A unity bit coding method by negative feedback," *Proc. IEEE*, vol. 51, pp. 1524-1533, November 1963.

[3.23] L. Williams, *Modeling and Design of High-Resolution Sigma-Delta Modulators*, Ph.D. Thesis, Stanford University, Technical Report No. ICL93-022, August 1993.

[3.24] B. Boser and B. A. Wooley, "The design of sigma-delta modulation analog-to-digital converters," *IEEE J. Solid-State Circuits,* vol. SC-23, pp. 1298-1308, December 1988.

[3.25] J. Candy, "A use of double integration in sigma delta modulation," *IEEE Trans. Commun.*, vol. COM-33, pp. 249-258, March 1985.

[3.26] B. P. Brandt, D. Wingard, and B. A. Wooley, "Second-order sigma-delta modulation for digital-audio signal acquisition," *IEEE J. Solid-State Circuits,* vol. SC-26, pp. 618-627, April 1991.

[3.27] K. Chao, S. Nadeem, W. Lee, and C. Sodini, "A higher order topology for interpolative modulators for oversampling A/D converters," *IEEE Trans. on Circuits and Systems II*, vol. CAS-37, pp. 309–318, March 1990.

[3.28] P. Ferguson, A. Ganesan, and R. Adams, "One bit higher order sigma-delta A/D converters," *Proc. 1990 IEEE Int. Symp. Circuits Syst.*, pp. 890-893, May 1990.

[3.29] T. Ritoniemi, T. Karema, and H. Tenhunen, "The design of stable high order 1-bit sigma-delta modulators," *Proc. 1990 IEEE Int. Symp. Circuits Syst.*, pp. 3267-3270, May 1990.

[3.30] B. DelSignore, D. Kerth, N. Sooch, and E. Swanson, "A monolithic 20-b delta-sigma A/D converter," *IEEE J. Solid-State Circuits*, vol. SC-25, pp. 1311–1317, December 1990.

[3.31] L. Williams and B. A. Wooley, "A third-order sigma-delta modulator with extended dynamic range," *IEEE J. Solid-State Circuits,* vol. SC-29, pp. 193-202, March 1994.

[3.32] Y. Matsuya, K. Uchimura, A. Iwata, T. Kobayashi, M. Ishikawa, and T. Yoshitome, "A 16-bit oversampling A-to-D conversion technology using triple-integration noise shaping," *IEEE J. Solid-State Circuits*, vol. SC-22, pp. 921–929, December 1987.

[3.33] M. Rebeschini, N. van Bavel, P. Rakers, R. Greene, J. Caldwell, and J. Haug, "A 16-b 160-kHz CMOS A/D converter using sigma-delta modulation," *IEEE J. Solid-State Circuits*, vol. 25, pp. 431-440, April 1990.

[3.34] L. Longo and M. Copeland, "A 13 bit ISDN-band oversampling ADC using two-stage third order noise shaping," *IEEE 1988 Custom Integrated Circuits Conference*, pp. 21.2.1–4, 1988.

[3.35] L. Williams and B. Wooley, "Third-order cascaded sigma-delta modulators," *IEEE Trans. on Circuits and Systems II*, vol. 38, pp. 489–498, May 1991.

[3.36] P. Gray and R. Meyer, *Analysis and Design of Analog Integrated Circuits*, John Wiley & Sons, 1993.

[3.37] R. Gray, "Quantization Noise Spectra," *Oversampled Delta-Sigma Data Converters Short Course*, Portland, Oregon, June 1992.

[3.38] B. P. Brandt and B. A. Wooley, "A 50-MHz multibit sigma-delta modulator for 12-b 2-MHz A/D conversion," *IEEE J. Solid-State Circuits,* vol. SC-26, pp. 1746-1756, December 1991.

[3.39] R. Baird and T. Fiez, "A low oversampling ratio 14-b 500-kHz DS ADC with a self-calibrated multibit DAC," *IEEE J. Solid-State Circuits,* vol. SC-31, pp. 312-320, March 1996.

[3.40] T. Cataltepe, G. Temes, and L. Larson, "Digitally corrected multibit $\Sigma\Delta$ data converters," *Proc. 1989 IEEE Int. Symp. Circuits Syst.*, pp. 647-650, May 1989.

[3.41] M. Sarhang-Nejad and G. Temes, "A high-resolution multibit $\Sigma\Delta$ ADC with digital correction and relaxed amplifier requirements," *IEEE J. Solid-State Circuits,* vol. SC-28, pp. 648-660, June 1993.

[3.42] C. Thompson and S. Bernadas, "A digitally-corrected 20b delta-sigma modulator," *ISSCC Digest of Tech. Papers*, pp. 194-195, February 1994.

[3.43] L. Carley, "A noise-shaping coder topology for 15+ bit converters," *IEEE J. Solid-State Circuits*, vol. SC-24, pp. 267-273, April 1989.

[3.44] J. Fattaruso, S. Kiriaki, M. de Wit, and G. Warwar, "Self-calibration techniques for a second-order multibit sigma-delta modulator," *IEEE J. Solid-State Circuits*, vol. SC-28, pp. 1216-1223, December 1993.

[3.45] Y. Sakina, "Multi-bit ΣΔ A/D converters with nonlinearity correction using dynamic barrel shifting," *University of California, Berkeley Report*, Dept. of EECS, 1990.

[3.46] F. Chen and B. Leung, "A high resolution multibit sigma-delta modulator with individual level averaging," *IEEE J. Solid-State Circuits*, vol. SC-30, pp. 453-460, April 1995.

[3.47] R. Baird and T. Fiez, "Linearity enhancement of multibit ΔΣ A/D and D/A converters using data weighted averaging," *IEEE Trans. on Circuits and Systems II*, vol. CAS-42, pp. 753-762, December 1995.

[3.48] T. Kwan, R. Adams, and R. Libert, "A stereo multibit sigma delta D/A with asynchronous master-clock interface," *IEEE J. Solid-State Circuits*, vol. SC-31, pp. 1881-1887, December 1996.

[3.49] H. Lin, J. Barreiro da Silva, B. Zhang, and R. Schreier, "Multi-bit DAC with noise-shaped element mismatch," *Proc. 1996 IEEE Int. Symp. Circuits Syst.*, pp. 235-238, May 1996.

[3.50] R. Henderson and O. Nys, "Dynamic element matching techniques with arbitrary noise shaping function," *Proc. 1996 IEEE Int. Symp. Circuits Syst.*, pp. 293-296, May 1996.

[3.51] I. Galton, "Spectral shaping of circuit errors in digital-to-analog converters," *IEEE Trans. on Circuits and Systems II*, vol. CAS-44, pp. 808-817, October 1997.

[3.52] L. Williams, "An audio DAC with 90dB linearity using MOS to metal-metal charge transfer," *ISSCC Digest of Tech. Papers*, pp. 58-59, February 1998.

CHAPTER 4   *Power Dissipation in Sigma-Delta A/D Converters*

This chapter examines the sources of power dissipation in a ΣΔ modulator and evaluates the relative merits of several circuits suitable for implementing low-power modulators that operate from a low supply voltage. The dominant source of power dissipation is identified as the first integrator in Section 4.1. Section 4.2 examines the power dissipation of several integrator architectures using ideal circuits. The impact of circuit nonidealities on the power dissipation of a switched-capacitor integrator is considered in Section 4.3, and the performance of several amplifier topologies is compared in Section 4.4. The chapter ends with a brief discussion of power dissipation in the decimation filter that is used to remove the out-of-band noise present in the modulator output.

4.1 Power Dissipation in a Sigma-Delta Modulator

Figure 4.1 is a simplified diagram of a single-stage ΣΔ modulator comprising an integrator followed by possible additional filtering and a quantizer. Gain in the forward path of the feedback loop desensitizes the modulator's output to subsequent disturbances and component nonidealities along the forward path. Errors within the low-frequency signal baseband are attenuated, while those at higher frequencies are amplified. This phenomenon, known as noise shaping, applies to circuit imperfections as well as the quantization error introduced by the quantizer. The order of shaping to

Power Dissipation in Sigma-Delta A/D Converters

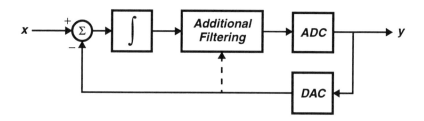

FIGURE 4.1 Sigma-delta modulator block diagram.

which a particular error in the forward path is subjected is equal to the order of filtering preceding the point in the circuit at which the error is introduced. Errors introduced at the input node are simply added to the signal and thus degrade the overall modulator performance directly.

Among the circuits is the forward path of a highly oversampled $\Sigma\Delta$ modulator, the first integrator stage has by far the greatest influence on the modulator's performance because of the noise shaping. Noise, settling, and harmonic distortion in this stage must generally meet the performance requirements specified for the overall A/D converter. Since power dissipation increases with increasing performance, this chapter focuses primarily on the design and implementation of the first integrator in a modulator of the type depicted in Figure 4.1.

Errors much greater than those introduced in the first stage can be tolerated in the subsequent filter and quantizer without significantly impairing a $\Sigma\Delta$ modulator's performance so long as the oversampling ratio is large enough. For example, with an oversampling ratio of 80, thermal noise introduced at the input to the first integrator is attenuated by a factor of 80 while that introduced at the output of the first integrator is reduced by a factor of 1.5×10^5 within the signal band. However, at low oversampling ratios the modulator becomes increasingly sensitive to noise in the stages following the first integrator. For example, with an oversampling ratio of only 4 thermal noise at the input to the first integrator is attenuated by a factor of 4 while thermal noise introduced at the output of the first integrator is reduced by a factor of only 19 in the signal band. In addition, noise attenuation can be further degraded if a subunity gain is introduced at the modulator input, as is discussed in Section 5.2. Fortunately, the techniques used to minimize the power dissipation in the first integrator are equally applicable to reducing the power in subsequent filtering stages.

4.2 Ideal Integrator Power Dissipation

This section examines three integrator implementations, switched-capacitor, continuous-time, and switched-current, in terms of their suitability for use as the first integrator stage in a power-efficient $\Sigma\Delta$ modulator operating from a low supply voltage. It is argued that the switched-capacitor approach provides the most advantageous trade-offs in low-voltage, low-power, high-linearity applications.

4.2.1 Switched-Capacitor Integrator

The processing of analog signals with a high degree of linearity has traditionally been accomplished using switched-capacitor techniques [4.1], [4.2]. Switched-capacitor circuits take advantage of the linear, well-matched capacitors available in CMOS technologies, which can be implemented using double-poly, poly-to-metal, or poly-to-diffusion structures or "sandwiches" of the back-end metal layers. Operational amplifiers together with the nearly ideal switches provided by MOS transistors make it possible to accurately transfer charge between capacitors. In recent work [4.3], a signal-to-distortion ratio of 110 dB over a 48-kHz bandwidth has been reported for a monolithic $\Sigma\Delta$ modulator implemented using switched-capacitor circuits.

The power consumed in a switched-capacitor integrator is generally proportional to its loading. Therefore, to minimize the power dissipation when such an integrator is used as the first stage of a $\Sigma\Delta$ modulator, the smallest capacitor sizes for which the required converter resolution and bandwidth can be maintained should be used. This approach leads to a modulator design for which the performance is limited primarily by kT/C noise in the first integrator.

Figure 4.2 illustrates a fully differential switched-capacitor integrator suitable for use as the first stage of the modulator in Figure 4.1. In the following equations, it is assumed that the operational amplifier has infinite bandwidth and gain, that there are no parasitic capacitances, and that the only noise in the integrator is that due to the sampling switches. The noise power within the baseband of an oversampling modulator that is introduced by a first stage integrator of the form shown in Figure 4.2 is then

$$S_{kT/C} = \frac{4kT}{MC_S}, \qquad (4.1)$$

where k is Boltzmann's constant, T is the absolute temperature, and M is the oversampling ratio, which is equal to the modulator sampling rate, f_S, divided by the Nyquist

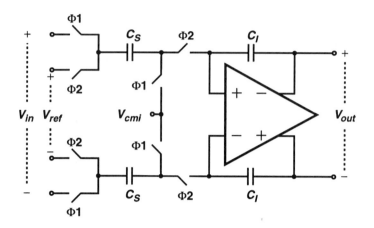

FIGURE 4.2 Switched-capacitor integrator.

sampling rate, f_N. The factor of 4 accounts for the two paths through which noise is sampled (during Φ1 and Φ2) and the fully differential structure of the circuit.

If the maximum amplitude of the differential input to the modulator is V_{sw}, then the power of a full-scale sinusoidal input is

$$S_S = \frac{V_{sw}^2}{2}. \qquad (4.2)$$

The dynamic range (DR) of the modulator is defined as the ratio of the power in a full-scale input to the power of a sinusoidal input for which the signal-to-noise ratio is one (0 dB). Thus, for a modulator whose baseband noise is dominated by kT/C noise in the first integrator stage, the dynamic range is

$$DR = \frac{S_S}{S_{kT/C}} = \frac{V_{sw}^2 M C_S}{8kT}. \qquad (4.3)$$

As the analysis presented in Appendix A.1 indicates, when implemented with an ideal class A amplifier, the minimum power dissipation in the first integrator stage can be estimated as

$$P = 32kT \times DR \times f_N, \qquad (4.4)$$

Ideal Integrator Power Dissipation

where DR is expressed as a ratio, rather than in dB, and f_N is the Nyquist sampling rate corresponding to the signal baseband. When implemented with an ideal class B amplifier, the power dissipation is

$$P = 4kT \times DR \times f_N \times \frac{\overline{\Delta V_{in}}}{V_{DD}}, \quad (4.5)$$

where $\overline{\Delta V_{in}}$ is the average of the difference between the sampled differential input, V_{in}, and the differential feedback reference voltage, V_{ref}, and V_{DD} is the supply voltage.

Equations (4.4) and (4.5) indicate that the minimum power dissipation in the first stage of an oversampling modulator depends linearly on the required dynamic range, DR. Since for every additional bit of resolution the DR must be increased by a factor of 4 (6 dB), the power dissipation depends very strongly on the resolution. The dependence on the Nyquist conversion rate, or equivalently the signal bandwidth, is linear. A key characteristic of the analytical results expressed in (4.4) and (4.5) is the absence of the oversampling ratio and, in the case of the class A amplifier, the supply voltage. Thus, the dependence of power dissipation on these factors is primarily an implementation issue.

4.2.2 Continuous-Time Integrator

An alternative to the switched-capacitor approach to realizing an analog integrator is continuous-time filtering. Figure 4.3 shows a continuous-time, two-input integrator wherein one input signal is applied through resistors and the other via a current source [4.4]. In the following equations, it is assumed that the amplifier has infinite bandwidth and gain, that its output can swing from rail-to-rail, that there are no parasitic capacitances, and that the only noise in the integrator is that due to the resistors. The spectrum of the voltage noise of a resistor is white, and its power spectral density is [4.5]

$$\frac{S_R}{\Delta f} = 4kTR. \quad (4.6)$$

Since in a $\Sigma\Delta$ modulator the resistor noise is not sampled until it has been filtered, the aliased components of the noise are attenuated by the noise shaping prior to sampling. Therefore, the input-referred noise power is approximately equal to the resistor noise

Power Dissipation in Sigma-Delta A/D Converters

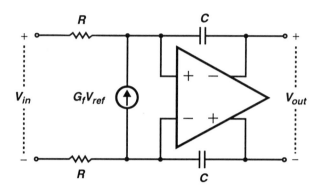

FIGURE 4.3 Continuous-time integrator.

that appears in the baseband, which, for the differential configuration of Figure 4.3, is given by

$$S_N = 8kTRf_N. \qquad (4.7)$$

If it is assumed that the power of the input signal is given by (4.2), the dynamic range, as defined in Section 4.2.1, is then

$$DR = \frac{S_S}{S_N} = \frac{V_{sw}^2}{16kTRf_N}. \qquad (4.8)$$

As the analysis presented in Appendix A.2 indicates, when implemented with an ideal class A amplifier, the minimum power dissipation in the integrator stage of Figure 4.3 can be estimated to be

$$P = 32kT \times DR \times f_N \times \frac{V_{DD}}{V_{sw}}, \qquad (4.9)$$

where DR is expressed as a ratio rather than in dB, and f_N is the Nyquist sampling rate corresponding to the signal baseband. When implemented with an ideal class B amplifier, the power dissipation is

Ideal Integrator Power Dissipation

$$P = 16kT \times DR \times f_N \times \frac{V_{DD}\overline{\Delta V_{in}}}{V_{sw}^2}, \qquad (4.10)$$

where $\overline{\Delta V_{in}}$ is the average difference between the differential input and feedback signals, $V_{in} - 2G_f RV_{ref}$, and V_{DD} is the supply voltage.

If off-chip resistors are used in a continuous-time integrator, the input signal amplitude is not constrained by the ESD input protection circuits and may exceed the supply rails, thereby reducing the power dissipation needed to achieve a specified level of performance. Furthermore, the amplifier settling requirements are generally more relaxed than in switched-capacitor circuits since charge transfer takes place uniformly over the entire clock period rather than being concentrated over a short fraction of the period. Thus, it is possible to obtain very high power efficiency with this approach [4.6], [4.7]. Unfortunately, there are many practical design issues that make it exceedingly difficult to achieve the dynamic range required for digital-audio applications with a continuous-time integrator [4.8], [4.9]. These include sensitivity to timing jitter, sensitivity to hysteresis in the feedback reference signal, and the need for either off-chip resistors or highly linear on-chip resistors. Because of these implementation difficulties in wide dynamic range applications, the continuous-time approach was not adopted in this work.

4.2.3 Switched-Current Integrator

In recent years, switched-current signal processing has been suggested as an alternative to the switched-capacitor approach because it avoids the need for linear capacitors and might be attractive for low-power and low-voltage implementations. However, as the following analysis indicates, the power dissipation of a switched-current integrator is actual higher that of its switched-capacitor counterpart. Moreover, it is difficult to achieve high linearity in a switched-current circuit when operating from a low supply voltage. Figure 4.4 illustrates a simple switched-current integrator [4.10]. In the following analysis, the MOS transistors are treated as square-law devices with infinite output resistance, and the switches are modeled as ideal switches in series with a resistor. Φ_1 and Φ_2, are two-phase, nonoverlapping clock signals that cause switches they control to conduct when they are high. The circuit operates as follows. During the phase when Φ_1 is high, the drain current of M_1 is

$$I_{D1}[n-1] = 2I_1 + I_{in}[n-1] - I_{D2}[n-1]. \qquad (4.11)$$

Power Dissipation in Sigma-Delta A/D Converters

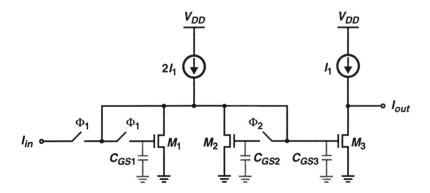

FIGURE 4.4 Switched-current integrator.

During the subsequent phases when Φ_2 is high and when Φ_1 is high again, the drain current of M_2 is

$$I_{D2}\left[n - \frac{1}{2}\right] = I_{D2}[n] = 2I_1 - I_{D1}[n-1]. \quad (4.12)$$

Since the drain current of M_3 is equal to that of M_2, the circuit implements the following transfer function

$$H(z) = \frac{z^{-1}}{1 - z^{-1}}, \quad (4.13)$$

which is the transfer function of a delaying integrator.

As in switched-capacitor integrators, the sampling operation introduces noise in this circuit. When Φ_1 falls, a noise voltage is sampled onto the gate of M_1 along with the desired gate voltage, with the voltage noise power given by

$$\overline{v_N^2} = \frac{kT}{C_{GS1}}, \quad (4.14)$$

where C_{GS1} is the gate to source capacitance of transistor M_1. The resulting drain current noise of M_1 is

Ideal Integrator Power Dissipation

$$\overline{i_N^2} = \frac{kT}{C_{GS1}} \times g_{m1}^2, \tag{4.15}$$

where g_{m1} is the transconductance of M_1.

The maximum input current, I_{in} in Figure 4.4, is limited by the drain current of M_1 when the voltage at its gate is at a maximum. Thus, the maximum current swing is achieved by biasing M_1 with a dc current equal to one half the maximum input current.

As the analysis in Appendix A.3 shows, the power dissipation of the circuit in Figure 4.4 is

$$P = 24kT(DR)f_N \frac{(V_{GS} - V_T)V_{DD}}{V_{in}^2}, \tag{4.16}$$

where V_{in} is the voltage swing on the gate of M_1. If V_{in} were allowed to approach V_{DD}, the power efficiency of switched-current circuits would approach that of continuous-time and switched-capacitor circuits. However, practical issues constrain V_{in} to a small fraction of V_{DD}, thereby degrading the circuit's power efficiency. In order to avoid gross nonlinearities that arise when there is a large voltage swing on the gate of M_1, V_{in} is typically constrained to less than 1/10 the overdrive voltage, $V_{GS} - V_T$, by limiting the input current swing. Moreover, at low supply voltages the threshold voltage of the MOS transistors becomes an increasing fraction of the supply voltage, further limiting the overdrive voltage and, hence, the input voltage swing.

Another significant source of nonlinearity in switched-current circuits is channel length modulation. To limit this effect, either cascode configurations or larger than minimum channel lengths must be used. Cascode current sources can be difficult to implement at low supply voltages, while long channel lengths limit the speed of the transistors. Another serious limitation of switched-current circuits is signal dependent charge injection from the MOS switches. Since the terminal voltages of the switches vary with input signal levels, channel charge injection when the switches are turned off also varies with signal level, giving rise to signal dependent errors. By contrast, the deleterious effects of signal dependent charge injection are eliminated in switched-capacitor circuits by taking advantage of the virtual ground terminal of an operational amplifier and by using delayed clock phases [4.2].

Power Dissipation in Sigma-Delta A/D Converters

Many enhancements over the circuit shown in Figure 4.4 have been reported in the literature. For example, [4.11] uses a 3-phase clock to reduce the effects of signal dependent charge injection in a 5-V filter, [4.12] describes a 3.3-V ΣΔ modulator using differential circuits, and [4.13] employs 142 pF of linear capacitance in each switched-current integrator to achieve better than 80-dB linearity in a 5-V ΣΔ modulator, at the cost of power, area, and speed. The reported performance of circuits that utilize switched-current techniques indicates that it has generally not been possible to implement low-power, low-voltage, high-resolution circuits with this approach.

4.3 Impact of Circuit Nonidealities

The power consumed by most practical integrators is typically three or more orders of magnitude greater than that predicted by the fundamental limits represented by (4.4) and (4.5). This is a consequence of the impact of various nonidealities, some of which are considered in this section as they relate to switched-capacitor circuits.

Equations (4.4) and (4.5) assume the use of ideal operational amplifiers and, thus, ignore the bandwidth and settling time of the integrator. Settling is one of the important factors increasing the power dissipation significantly above the levels predicted by (4.4) and (4.5). In response to a step input, the integrator output typically slews for some time and then enters a linear settling regime. In high-resolution applications, linear settling to the full resolution of the converter is usually required. Therefore, the integrator settling time is substantially greater than it would be if the integrators were only slew limited. Consequently, the power dissipation for a given f_N is increased relative to that predicted by (4.4) and (4.5).

In many operational amplifier circuits, the settling time constant is a function of the size of the input transistors. Consider an amplifier in the feedback configuration shown in Figure 4.5. This configuration corresponds to a switched-capacitor integrator during the charge transfer phase. C_S is the sampling capacitor, C_I is the integration capacitor, C_P is the input capacitance, which is dominated by the amplifier's input transistors, and C_L is the load capacitance, which includes the bottom plate parasitic of C_I as well as loading from the subsequent stage. Assume that the open loop transfer function of the amplifier may be modeled as a single pole response of the form

$$A(s) = \frac{A_0}{1 - s/p_1}, \quad (4.17)$$

Impact of Circuit Nonidealities

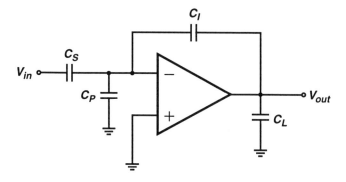

FIGURE 4.5 Operational amplifier with capacitive feedback.

where A_0 is the open loop dc gain and p_1 is the open loop pole. Further assume that the amplifier dc gain is

$$A_0 = g_m r_{out}, \quad (4.18)$$

where g_m is the transconductance of the input transistor and r_{out} is the open loop output resistance of the amplifier, and that the pole in the amplifier response is

$$p_1 = -\frac{1}{r_{out} C_{Leff}}, \quad (4.19)$$

where C_{Leff} is the effective capacitive load. It is straightforward to show that, if the amplifier is modeled using (4.17)-(4.19), the settling time constant of the circuit in Figure 4.5 is

$$\tau = \frac{C_L + C_S + C_P + (C_L/C_I)(C_S + C_P)}{g_m}. \quad (4.20)$$

It can be seen from (4.20) that increasing the size of the input transistor gives rise to two competing effects: a reduction in the settling time by increasing g_m, but also an increase in the settling time as a result of the increase in C_P. Since C_P increases faster than g_m, there is an input transistor size that minimizes τ. This optimum size is a function of the capacitances in the circuit, as well as the degree to which the input transistors are velocity saturated. Thus, the impact of the input capacitance is to increase the amplifier settling time, and the dependence of the input capacitance on the input transistor size limits the extent to which increasing the input transistor width can shorten the response time.

If the oversampling ratio and f_N of a modulator are held constant and its dynamic range is to be increased, then the size of the sampling capacitors in the first stage must be increased. This allows for the use of larger input transistors, which thereby diminishes the relative impact of switch and wiring parasitics on the settling response. Thus, the power dissipation in the modulator moves closer to its fundamental limit as the resolution is increased. On the other hand, if the oversampling ratio and dynamic range of a modulator are held constant while f_N is increased, the settling time must be decreased proportionally to maintain the required settling precision. If the first-stage integrator is implemented with a class A amplifier, then although the slew rate increases linearly with tail current, the settling time constant only decreases in proportion to the square root of the tail current. Therefore, the fraction of the settling time devoted to linear settling becomes larger as the tail current is increased, and the power dissipation of the converter increases relative to the level predicted by the fundamental limit. Thus, for a class A amplifier, the power dissipation transitions from a linear to a quadratic dependence on f_N, rather than following the simple linear dependence predicted by (4.4) and (4.5).

Another important practical deviation from the ideal equations is the limited amplifier output swing. To maintain a constant *DR* as the supply voltage is lowered, the noise power must be reduced by the same amount that the signal power is reduced. This implies that capacitor sizes, and hence amplifier currents, must be increased by the same proportion that signal power is reduced. Since signal swing scales proportionally faster than supply voltage, there is a net increase in power dissipation as supply voltage is reduced, as is apparent in equations (A.8) and (A.13) in Appendix A.1.

While equations (4.4) and (4.5) take into account only *kT/C* noise arising from the sampling network, the operational amplifier also contributes noise within the baseband. The two dominant types of amplifier noise in digital-audio applications are flicker (*1/f*) noise and broadband thermal noise. Several circuit techniques are commonly used to suppress *1/f* noise: chopper stabilization, correlated double sampling, and transistor sizing. However, broadband noise cannot be greatly reduced in a power efficient manner. Typically, the best noise performance is obtained with a simple differential-pair input stage since folded structures introduce additional noise sources. When the amplifier thermal noise is dominated by the noise of a differential-pair with a current mirror load, the baseband input-referred thermal noise of the first integrator is

Impact of Circuit Nonidealities

$$\overline{v_N^2} = \frac{2kT}{MC_C} \times \frac{\gamma}{1 + C_S/C_I} \times \left(1 + \frac{g_{m3}}{g_{m1}}\right), \quad (4.21)$$

where C_C is the amplifier compensation capacitor, γ is the noise enhancement factor of short channel transistors [4.14], g_{m1} is the transconductance of the input transistors, and g_{m3} is the transconductance of the load transistors. Increasing the compensation capacitor reduces the amplifier thermal noise but requires higher currents to obtain the desired bandwidth. A reasonable compromise is achieved by choosing C_C equal to C_S.

A significant simplification in Section 4.2.1 is the omission of parasitic capacitances. As mentioned earlier, the amplifier input capacitance can limit the settling response of the integrator. Another important consideration is the bottom plate parasitic of the integration capacitor, which directly loads the amplifier output. Using (4.20) it can be shown that the relative increase in the settling time constant as a function of the bottom plate parasitic of C_I is

$$\frac{\tau(C_L)}{\tau(C_L = 0)} = 1 + \frac{C_L}{C_I}\left(\frac{C_S + C_P + C_I}{C_S + C_P}\right). \quad (4.22)$$

For example, when $C_S = 4\,\text{pF}$, $C_P = 1\,\text{pF}$, $C_I = 20\,\text{pF}$, and $C_L = 4\,\text{pF}$, which is typical of a poly-to-poly analog capacitor with a 20% bottom plate parasitic, the increase in the settling time constant is two fold in comparison with $C_L = 0$. If $C_L = 16\,\text{pF}$, which is typical of back-end metal sandwich capacitors with an 80% bottom plate parasitic, the settling time constant is increased by a factor of five. The increase in τ, in turn, necessitates a commensurate increase in power dissipation to maintain the required settling response.

Equations (4.4) and (4.5) show no dependence of the power dissipation on the modulator oversampling ratio, M. The implicit assumption is that, as M is increased, the capacitor sizes can be decreased in proportion to maintain constant noise levels, as indicated by (4.1). Therefore, the settling response can be held approximately constant without an increase in power dissipation. In practice, matching considerations can limit the scaling of capacitor sizes, while the contribution of parasitic capacitances, particularly the amplifier input capacitance, becomes more significant as the sampling capacitor size is reduced. This, in turn, limits the extent to which the size of the input transistors can be increased, thus requiring higher currents to maintain the settling response as M is increased.

The idealized amplifiers assumed in Section 4.2 have no ancillary current paths. In practice, amplifier structures require additional circuitry and current for biasing and common-mode feedback. It is therefore important to use circuits for these functions that require as little current as possible, employing techniques such as switched-capacitor common-mode feedback so as to avoid the use of additional differential pairs.

Additional practical considerations are the degradation in settling time that can result from the feedforward zero of the switched-capacitor integrator during the integration phase, as well as noise coupling through the supplies, substrate, and bias lines that can increase the baseband noise floor and introduce harmonic distortion.

4.4 Comparison of Amplifier Topologies

The choice of amplifier topology plays a critical role in low-voltage, low-power integrator design. The merits of three amplifier topologies are examined here: folded cascode, two-stage class A, and two-stage class A/AB. Simplified circuit schematics for these configurations are presented in Figures 4.6(a), (b), and (c), respectively. Table 4.1 summarizes the approximate dependence of key performance metrics on various device and circuit parameters. The folded cascode topology has the highest nondominant pole, and thus provides the highest frequency performance. However, it also has the lowest output signal swing and is somewhat noisier than the other circuits.

The two-stage class A amplifier in Figure 4.6(b) has a lower nondominant pole than the folded cascode circuit. Thus, in high-speed applications the two-stage class A circuit requires relatively high currents in the second stage to push out the lowest nondominant pole, thereby significantly increasing its power consumption. A further shortcoming of the two-stage class A amplifier is its power supply rejection ratio (PSRR) from the bottom rail, which is particularly poor at high frequencies. Supply noise gives rise to a common-mode output signal in the fully differential circuit that can be translated into a differential-mode error in the presence of device mismatch. Biasing of the two-stage class A amplifier is typically accomplished with a common-mode feedback circuit that senses the output common-mode voltage in order to control the tail current source via a current mirror. Owing to stability considerations the gain and bandwidth of the common-mode feedback loop are limited to at most those of the differential mode signal path.

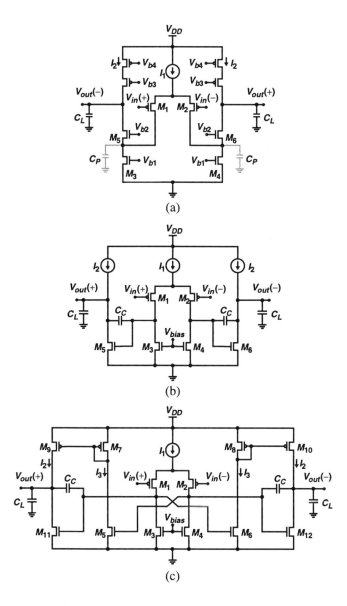

FIGURE 4.6 Amplifiers: (a) folded cascode, (b) two-stage class A, (c) two-stage class A/AB.

Power Dissipation in Sigma-Delta A/D Converters

TABLE 4.1 Key parameters of three amplifier topologies.

Parameter	Folded Cascode	Two-Stage Class A	Two-Stage Class A/AB
Slew Rate	I_1/C_L	min $(I_1/C_C, 2I_2/(C_L+C_C))$	I_1/C_C
Unity Gain Freq, ω_u	g_{m1}/C_L	g_{m1}/C_C	g_{m1}/C_C
Lowest Non-dominant pole, ω_2	g_{m5}/C_P	g_{m5}/C_L	$2g_{m11}/C_L$
Output Swing	$2V_{DD}-8|V_{DS,min}|$	$2V_{DD}-4|V_{DS,min}|$	$2V_{DD}-4|V_{DS,min}|$
Thermal Noise	$8\gamma kT/g_{m1} \times (1+g_{m3}/g_{m1}+g_{m5}/g_{m1})$	$8\gamma kT/g_{m1} \times (1+g_{m3}/g_{m1})$	$8\gamma kT/g_{m1} \times (1+g_{m3}/g_{m1})$
Minimum Supply	$|V_{T1}|+2|V_{DS,min}|$	$|V_{T1}|+2|V_{DS,min}|$	$|V_{T1}|+2|V_{DS,min}|$
CMRR	+	+	++
PSRR	+	−	++

The two-stage class A/AB amplifier of Figure 4.6(c) combines a simple differential-pair as the first stage with a class A/B second stage wherein push-pull operation is implemented through the use of current mirrors. Slew limiting only occurs in the first stage. The second-stage currents are chosen so that the nondominant poles are sufficiently high in frequency to ensure stability. Because of the push-pull operation, the lowest nondominant pole in the class A/AB design is governed by the time constant formed by approximately twice the transconductance of the output NMOS transistor and the load capacitance. Thus, the output branch current can be about half that used in the two-stage class A circuit for the same nondominant pole frequency. When this fact is exploited together with the use of gain in the second-stage current mirrors, a significant reduction in power dissipation can be achieved relative to the two-stage class A topology. Increasing the gain in the current mirrors does lower the mirror pole and will eventually degrade the phase margin of the circuit. Therefore, the power dissipation advantage of this circuit is only attained in relatively low-speed applications. Even with a current mirror gain of one, however, the power dissipation of the circuit is at least comparable to that of the two-stage class A design.

A consequence of using current mirrors in the second stage of the class A/AB design is that the common-mode output voltage of the first stage influences the bias currents in the second stage but does not control the second-stage output voltage. Thus, it is not possible to stabilize the common-mode levels in the circuit with a single feedback loop. The common-mode voltages of the two stages can be established with high precision through the use of two independent common-mode feedback loops with bandwidths that can exceed the bandwidth of the differential-mode signal path. Noise on the bottom supply rail appears as a common-mode input to the second stage and is thus attenuated by the common-mode rejection of that stage. Consequently, the PSRR of this circuit is superior to that of the two-stage class A amplifier.

The minimum power supply voltage from which the three different amplifiers can operate is governed by the input stage if the circuits are implemented in a technology wherein only transistors with conventional high threshold voltages are available. In that case, all three amplifiers have the same minimum supply level, which is equal to sum of the common-mode input level, the gate-to-source voltage of the input transistors, and the saturation voltage of the tail current source. In switched-capacitor integrators of the form shown in Figure 4.2, the input and output common-mode voltages can be set independently. Therefore, by using a low common-mode input voltage of 400 mV and saturation voltages of 150 mV, integrators implemented in a technology with 800-mV transistor thresholds can be operated from supply voltages as low as 1.5 V. If the minimum supply voltage is governed by the output stage, as may be the case for amplifiers realized in a technology in which low threshold voltage transistors are available, the two-stage circuits can operate from lower supply levels than the folded cascode amplifier.

Figure 4.7 plots hand-calculated estimates, derived in Appendix B, of the power dissipated in the first-integrator of an oversampling modulator as a function of supply voltage. The power dissipation in all three circuits increases sharply at low supply voltages. However, the larger output swing of the two-stage amplifiers does make them more suitable for low-voltage operation. The deleterious impact of low-voltage operation on the power dissipation can be mitigated by using a lower $V_{DS,min}$. This approach was adopted in [4.15], where $V_{DS,min}$ equal to 50 mV was used in a telescopic cascode amplifier to achieve 1.5-V operation at the cost of speed and linearity. Unfortunately, the intrinsic speed of CMOS transistors is proportional to the transistor overdrive voltage [4.5]. Therefore, a very low $V_{DS,min}$ is only practical in low-speed applications.

Figure 4.8 plots hand-calculated estimates, derived in Appendix B, of power dissipation as function of oversampling ratio for all three amplifier topologies. At low oversampling ratios, the power dissipation is essentially independent of the oversampling ratio, while at high oversampling ratios the power dissipation rises rapidly with the oversampling ratio. The increase in power consumption at high oversampling ratios is due to the increased impact of parasitic capacitances as the sampling and integration capacitors are reduced in size. Based on the results presented in Figure 4.8, an over-

Power Dissipation in Sigma-Delta A/D Converters

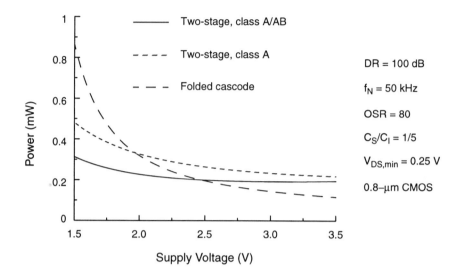

FIGURE 4.7 First integrator power dissipation vs. supply voltage.

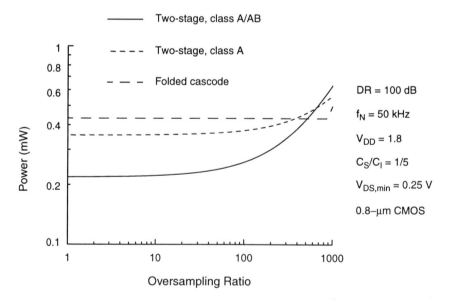

FIGURE 4.8 First integrator power dissipation vs. oversampling ratio.

sampling ratio of 80 appears suitable for the realization of a low-power modulator that meets the performance demands of digital-audio applications, ensuring operation in the low-speed regime while retaining the benefits of a reasonably high oversampling ratio. A two-stage class A/AB amplifier has been used in the experimental prototype described in the following chapters because of its low power dissipation, its good control of common-mode levels, and its rejection of power supply noise.

4.5 Power Dissipation in the Decimation Filter

An A/D converter based on $\Sigma\Delta$ modulation combines an oversampling modulator that shapes the quantization noise with a digital decimation filter that removes the shaped out-of-band noise and lowers the sampling rate of the output data to the Nyquist rate. The power dissipated in the decimation filter is a function of the filter specifications, architecture and fabrication technology. Although the filter power consumption is typically less than that of the modulator, it nevertheless represents a significant portion of the overall converter power dissipation. In this section, some published work is reviewed to illustrate the relative impact of the digital and analog portions of a $\Sigma\Delta$ oversampling A/D converter on power dissipation.

In [4.16], dual $\Sigma\Delta$ digital-audio band modulators and decimation filters are fabricated in a 1.2-µm BiCMOS technology. When operating from a 3-V power supply, the converter achieves a 93-dB *SNR*. The decimation filter is implemented as six sinc filter stages followed by two halfband FIR filters of lengths 31 and 167. The filter, which comprises 88000 transistors, provides 120 dB of out-of-band attenuation and a passband ripple of 3×10^{-4} dB. The power dissipation of each filter at 3 V is 15 mW, while each modulator dissipates 27 mW.

[4.17] describes a decimation filter suitable for digital-audio band $\Sigma\Delta$ modulators. This circuit was fabricated in a standard 1-µm CMOS technology and can operate from a 3-V supply. The filter is a cascade of a third-order sinc filter followed by two halfband FIR filter of lengths 18 and 110 and a droop correction FIR filter of length 8. The out-of-band attenuation is 80-dB and the passband ripple is less than 1×10^{-2} dB. The power dissipation of this filter at 3 V is 6.5 mW. The power dissipation reported for a modulator implemented in this technology is 14 mW [4.18].

Another digital-audio decimation filter is described in [4.19]. This circuit was fabricated in a 0.7-µm CMOS process with low threshold voltage devices. The architecture is a merged multistage sinc filter followed by a 135 tap FIR polyphase filter. Much of the filter hardware is shared among dual $\Sigma\Delta$ modulators. The out-of-band attenuation is 110-dB and the passband ripple is 1×10^{-3} dB. The power dissipation of this filter per output channel is 10 mW from a 3.3-V supply. Each modulator dissipates 250 mW.

The results cited above indicate that the power dissipation of digital decimation filters suitable for use with digital-audio $\Sigma\Delta$ modulators is a relatively low, but significant, portion of the overall converter power dissipation, even in a 1-μm technology. However, the power dissipation in the digital filters will fall quickly with the scaling of technology and supply voltage, while the power dissipated in the analog circuits is likely to increase (see Section 2.2 and Section 4.3). Therefore, techniques for reducing the power dissipation in the modulator are essential to achieving an overall reduction in the power consumed by an oversampling A/D converter.

4.6 Summary

This chapter has examined the dominant contributor to power dissipation in a $\Sigma\Delta$ modulator, the first integrator stage. The suitability of several integrator architectures for low-power and low-voltage applications has been evaluated, and switched-capacitor circuits have been deemed a favorable choice for high-resolution applications. Among amplifier circuits that can be used to implement a switched-capacitor integrator, a high-swing, two-stage class A/AB topology appears to be especially attractive for use in a supply-constrained environment. Finally, the contribution of digital decimation filters to the overall power dissipation of a $\Sigma\Delta$ oversampling A/D converter has been reviewed. The relatively low power dissipation of the digital filter will continue to fall as technology scales, while power reduction in the modulator will require further advances in both architecture and circuit design.

REFERENCES

[4.1] K. Hsieh, P. Gray, D. Senderowicz, and D. Messerschmitt, "A low-noise chopper-stabilized differential switched-capacitor filtering technique," *IEEE J. Solid-State Circuits*, vol. SC-16, pp. 708–715, December 1981.

[4.2] K. L. Lee and R. Meyer, "Low-distortion switched-capacitor filter design techniques," *IEEE J. Solid-State Circuits,* vol. SC-20, pp. 1103-1112, December 1985.

[4.3] K. Leung, E. Swanson, K. Leung, and S. Zhu, "A 5V, 118dB ΣΔ analog-to-digital converter for wideband digital audio," *ISSCC Digest of Tech. Papers*, pp. 218-219, February 1997.

[4.4] B. DelSignore, D. Kerth, N. Sooch, and E. Swanson, "A monolithic 20-b delta-sigma A/D converter," *IEEE J. Solid-State Circuits*, vol. SC-25, pp. 1311–1317, December 1990.

[4.5] P. Gray and R. Meyer, *Analysis and Design of Analog Integrated Circuits*, John Wiley & Sons, 1993.

[4.6] E. J. van der Zwan and E. C. Dijkmans, "A 0.2mW CMOS ΣΔ modulator for speech coding with 80dB dynamic range," *ISSCC Digest of Tech. Papers*, pp. 232-233, February 1996.

[4.7] E. J. van der Zwan, "A 2.3mW CMOS ΣΔ modulator for audio applications," *ISSCC Digest of Tech. Papers*, pp. 220-221, February 1997.

[4.8] L. Williams, *Modeling and Design of High-Resolution Sigma-Delta Modulators*, Ph.D. Thesis, Stanford University, Technical Report No. ICL93-022, August 1993.

[4.9] E. J. van der Zwan and E. C. Dijkmans, "A 0.2mW CMOS ΣΔ modulator for speech coding with 80dB dynamic range," *IEEE J. Solid-State Circuits,* vol. SC-31, pp. 1873-1880, December 1996.

[4.10] T. Fiez, G. Liang, and D. Allstot, "Switched-current circuit design issues," *IEEE J. Solid-State Circuits,* vol. SC-26, pp. 192-202, March 1991.

[4.11] J. Hughes and K. Moulding, "An 8MHz, 80Ms/s switched-current filter," *ISSCC Digest of Tech. Papers*, pp. 60-61, February 1994.

[4.12] N. Tan and S. Eriksson, "A low-voltage switched-current sigma-delta modulator," *IEEE J. Solid-State Circuits,* vol. SC-30, pp. 599-603, May 1995.

[4.13] N. Moeneclaey and A. Kaiser, "Design techniques for high-resolution current-mode sigma-delta modulators," *IEEE J. Solid-State Circuits,* vol. SC-32, pp. 953-958, July 1997.

[4.14] A. Abidi, "High frequency noise measurements on FET's with small dimensions," *IEEE Trans. on Electron Devices,* vol. ED-33, pp. 1801-1805, November 1986.

[4.15] D. Senderowicz, et al., "Low-voltage double-sampled converters," *ISSCC Digest of Tech. Papers,* pp. 210-211, February 1997.

[4.16] T. Ritoniemi, E. Pajarre, S. Ingalsuo, T. Husu, V. Eerola, T. Saramäki, "A stereo audio sigma-delta A/D-converter," *IEEE J. Solid-State Circuits,* vol. SC-29, pp. 1514-23, December 1994.

[4.17] B. P. Brandt and B. A. Wooley, "A low-power, area-efficient digital filter for decimation and interpolation," *IEEE J. Solid-State Circuits,* vol. SC-29, pp. 679-687, June 1994.

[4.18] B. P. Brandt, D. Wingard, and B. A. Wooley, "Second-order sigma-delta modulation for digital-audio signal acquisition," *IEEE J. Solid-State Circuits,* vol. SC-26, pp. 618-627, April 1991.

[4.19] I. Fujimori, et al., "A 5-V single chip delta-sigma audio A/D converter with 111 dB dynamic range," *IEEE J. Solid-State Circuits,* vol. SC-32, pp. 329-336, March 1997.

CHAPTER 5

Design of a Low-Voltage, High-Resolution Sigma-Delta Modulator

This chapter discusses the architecture and circuit requirements for a CMOS $\Sigma\Delta$ modulator that provides digital-audio performance when operated from a supply voltage of less than 2 V. The primary performance objective is to achieve a dynamic range of 98-dB (16-bit) for a 25-kHz signal bandwidth, while operating from a single 1.8-V supply with the lowest possible power dissipation. The following chapter describes the circuits used to implement this modulator in a 0.8-µm, double-metal, CMOS technology with poly-to-metal capacitors.

As shown in the previous chapter, an oversampling ratio as high as 80 does not appreciably increase the power dissipation of a digital-audio band $\Sigma\Delta$ modulator integrated in 0.8-µm CMOS. This chapter opens with a comparison of three architectures that can deliver digital-audio performance with an oversampling ratio of 80. The comparison is followed by a discussion of scaling the internal signal swings in a 2-1 cascaded modulator to make low-voltage operation feasible. The impact of implementation nonidealities, including capacitor mismatch, integrator settling, and circuit noise, is then considered. The chapter concludes with specifications for the circuits used to implement an experimental $\Sigma\Delta$ modulator, while the design of these circuits is described in Chapter 6.

Design of a Low-Voltage, High-Resolution Sigma-Delta Modulator

5.1 Modulator Architecture

Based on the power dissipation analysis of the previous chapter, an oversampling ratio of 80 has been chosen for the experimental modulator. The primary objective in selecting an architecture is to reduce the quantization noise so that the total modulator noise floor of $-100\,\text{dB}$ is dominated by circuit noise.

With an oversampling ratio of 80, third-order noise shaping is needed to attenuate the quantization noise to a level acceptable for digital-audio applications when using a 1-bit quantizer. A 2-1 cascaded architecture provides performance close to that of an ideal third-order noise-differencing modulator and is, therefore, a suitable candidate. Alternatively, a single-stage modulator may be used. Single-stage modulators of order higher than two must be stabilized by restricting the range of the input signal and by tailoring the noise transfer function to reduce the quantization noise power at high frequencies at the cost of higher baseband noise [5.1]. For these reasons, high-order single-stage modulators generally have lower dynamic range than cascaded modulators of the same order. A single-stage architecture that provides digital-audio performance at an oversampling ratio of 80 is the fourth-order modulator described in [5.2]. Another architecture that provides adequate performance for digital-audio applications is a second-order modulator employing a 4-bit quantizer. The quantization noise floor, overload level relative to a full-scale input, spectral tones and circuit requirements of these architectures are compared in the following subsections.

5.1.1 Third-Order (2-1) Cascaded Modulator

The block diagram of a 2-1 cascaded modulator shown in Figure 3.15 is repeated in Figure 5.1. The transfer function, dynamic range, and sensitivity to mismatch and integrator leak in the 2-1 cascaded architecture were analyzed in Section 3.3.5 under the white noise approximation. In this section, computer simulations that model the nonlinearity of the quantizer are used to select the system gains, b, β, and λ. The values of these gains are chosen with the objective of maximizing the dynamic range of the modulator under the assumption that the circuit noise floor is $-100\,\text{dB}$. This amounts to maximizing the modulator overload level without increasing the quantization noise floor above the circuit noise. Recall that in this text an input level of $0\,\text{dB}$ corresponds to the power in an input sinusoid with a peak-to-peak amplitude equal to the feedback reference voltage (adjusted for any difference in the gains at the inputs of the first integrator stage). Another consideration in choosing the system gains is to ensure that the baseband quantization noise is independent of the input signal, thereby avoiding the presence of spurious tones and signal dependence in the noise floor.

Modulator Architecture

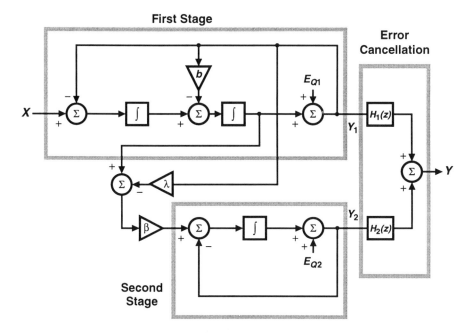

FIGURE 5.1 Block diagram of a 2-1 cascaded sigma-delta modulator.

As shown in Section 3.3.5, the quantization noise of the first stage is removed by means of an error cancellation network in a 2-1 cascaded architecture. It is nevertheless important to achieve good performance in the first stage so that, in the presence of component mismatch and integrator leak, imperfect cancellation of first-stage quantization noise will not significantly degrade the modulator's performance. Therefore, the choice of the feedback coefficient, b, in the first-stage of the modulator of Figure 5.1 is a significant consideration.

The simulated baseband *SNDR* of the first stage of the modulator of Figure 5.1 is shown in Figure 5.2 as a function of the input signal power for several values of b. Under the assumption of a Nyquist sampling rate of 50 kHz, an input sinusoid at 1.01 kHz was used to generate the data in Figure 5.2 and subsequent *SNDR* plots in this chapter. It is apparent from Figure 5.2 that, while the quantization noise floor is insensitive to b for the range of values simulated, the peak *SNDR*, the overload level, and signal dependence in the quantization noise are all affected by the choice of b. The peak *SNDR* and the overload level are highest when $2 < b < 4$. Moreover, the

Design of a Low-Voltage, High-Resolution Sigma-Delta Modulator

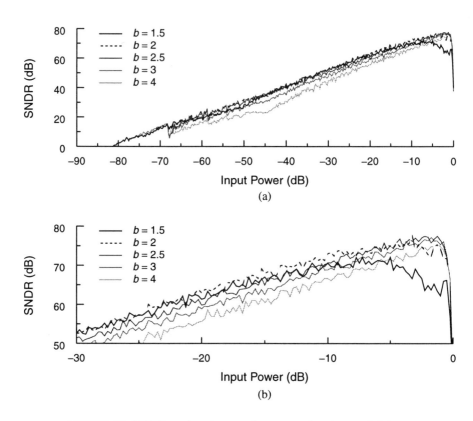

FIGURE 5.2 SNDR vs. input power in a second-order modulator.

quantization noise exhibits greater signal dependence when $b \geq 3$. For this reason, a value of $b = 2.5$ was used in the experimental prototype.

A consequence of choosing $b = 2.5$ is a reduction in spectral tones that was noted in [5.3]. When $b = 2$, strong tones appear near the zero input level, where they can be especially objectionable since the signal power is low. When $b = 2.5$, the spectral tones occur at higher input powers and are reduced in amplitude.

As shown in the analysis of Section 3.3.3, b must be equal to 2 for the second-order system to implement a noise-differencing transfer function. However, the spectral tone and overload performance of the first stage of the modulator in Figure 5.1 with $b = 2.5$ are superior to those of the true noise-differencing architecture.

Modulator Architecture

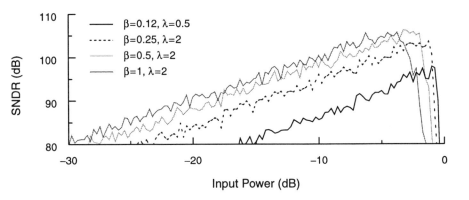

FIGURE 5.3 SNDR vs. input power in a 2-1 cascaded modulator.

Linear analysis using the root-locus method and based on the white noise and unity gain approximations predicts that a second-order modulator with a feedback coefficient greater than or equal to 2.5 is unstable. However, the simulation results plotted in Figure 5.2 clearly indicate that this is not the case. Careful analysis using the method of describing functions has shown that this discrepancy is due to the fact that the equivalent gain of the 1-bit quantizer is a function of the input signal as well as the feedback coefficient, decreasing as b is increased, so that the system remains stable even for large b [5.4], [5.5].

In Section 3.3.5, linear approximations were used to show that the quantization noise power in a 2-1 cascaded modulator is proportional to $1/\beta^2$, as is evident in (3.49). Further investigation using simulations that model the nonlinear behavior of the quantizer indicates that the choice of both β and λ significantly impacts the modulator's overload level and peak *SNDR*, as well as its quantization noise floor. Figure 5.3 shows the *SNDR* of the modulator of Figure 5.1 as a function of the input signal power for several values of β and λ with $b = 2.5$. To generate the data for Figure 5.3, λ was swept for each value of β shown, and the data for the value of λ that generated the highest overload level was retained. The highest overload level is −0.65 dB, which is observed when $\beta = 0.12$ and $\lambda = 0.5$. Unfortunately, this combination of gains increases the quantization noise floor above the performance target for the experimental modulator. The next highest overload level of −1.0 dB is achieved when $\beta = 0.25$ and $\lambda = 2$. Since the quantization noise floor for this combination of gains is −105 dB, this architecture is a candidate for use when the thermal noise floor is expected to be −100 dB or higher. The two other combinations of β and λ have lower

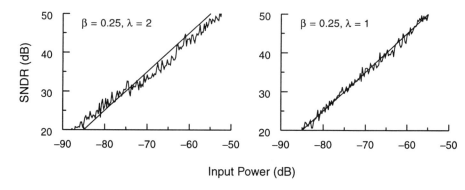

FIGURE 5.4 SNDR vs. input power showing signal dependence in noise floor.

noise floors, but also lower overload levels. In fact, in the absence of thermal noise the dynamic range of the modulator is maximized when $\beta = 0.5$ and $\lambda = 2$.

In the experimental prototype, gains of $\beta = 0.25$ and $\lambda = 1$ were used, although they yield a slightly lower overload level of $-1.3\,\text{dB}$ than when $\beta = 0.25$ and $\lambda = 2$. This decision was driven by the fact that the selected combination of gains results in a quantization noise floor that is very nearly independent of signal power. This can be seen in Figure 5.4 where the *SNDR* is plotted as a function of the input signal power for $\beta = 0.25$ and $\lambda = 2$, as well as $\beta = 0.25$ and $\lambda = 1$, along with a straight line whose slope is equal to 1. There is little observable fluctuation in baseband quantization noise with the latter combination of gains.

To investigate the occurrence of spectral tones in a 2-1 cascade with $b = 2.5$, $\beta = 0.25$, and $\lambda = 1$, a simulation was performed wherein a dc input signal was swept in fine increments across the positive input range of the modulator with the feedback reference levels equal to ± 0.5. At each input level, the modulator output was decimated to the Nyquist rate, an FFT was performed, and the power of the strongest spectral component in the baseband was recorded. The results of this simulation can be see in Figure 5.5. It is apparent that there are no tones whose power exceeds $-100\,\text{dB}$ when the dc input is less than 0.416.

5.1.2 Fourth-Order Single Stage Modulator

Another modulator architecture suitable for obtaining digital-audio performance is the single-stage, fourth-order topology described in [5.2]. This architecture employs a

Modulator Architecture

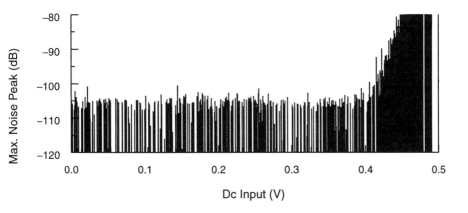

FIGURE 5.5 Spectral tones in a 2-1 cascaded modulator.

forward path filter that generates an equiripple baseband quantization noise transfer function and prevents the magnitude of the out-of-band response from growing so large as to cause instability.

Figure 5.6 plots the *SNDR* as a function of the input signal power for the modulator of [5.2] when the oversampling ratio is 80. The overload level is −3.7 dB, the noise floor is −108 dB, and the peak *SNDR* is only 96 dB. An increase in the quantization noise can be observed at an input signal power of −75 dB. It has been shown that the output of this modulator contains strong spectral tones at low input signal power levels [5.3].

Functional simulations of circuit nonidealities indicate [5.2] that capacitor matching on the order of 5% and amplifier gain on the order of 200 are needed in order to prevent the baseband quantization noise from increasing significantly above that achieved with ideal circuits. Furthermore, the integrator settling requirements in this single-stage, fourth-order system are similar to those in the cascaded modulator. Thus, the circuit and matching requirements of this architecture are approximately equivalent to those in the cascaded modulator when the oversampling ratio is 80.

5.1.3 Second-Order Modulator with Multibit Quantization

A third architecture that achieves the performance needed for digital-audio applications with an oversampling ratio of 80 is a second-order modulator with a 4-bit quantizer. The use of a multibit quantizer makes linear analysis more representative of the

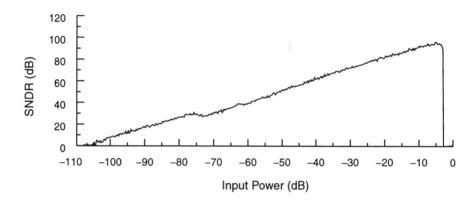

FIGURE 5.6 SNDR vs. input power in a fourth-order, single-stage modulator.

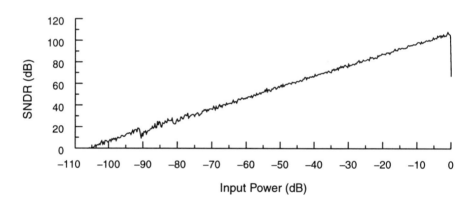

FIGURE 5.7 SNDR vs. input power in a 4-bit, second-order modulator.

modulator's behavior since the quantization noise of a multibit quantizer more nearly resembles the white noise approximation than that of a 1-bit quantizer, and the quantizer's gain is well defined. Thus, there is no need to invoke the unity gain approximation when analyzing a modulator employing a multibit quantizer.

Figure 5.7 shows the *SNDR* as a function of the input signal power for a second-order, 4-bit modulator with an oversampling ratio of 80. The overload level is −0.2 dB, the noise floor is −106 dB, and the peak *SNDR* is 107 dB. Thus, the overload level is the highest among the architectures examined here. However, at low input signal levels, the noise floor exhibits some signal dependence.

An important advantage of using a multibit quantizer is that the integrator settling requirements are greatly relaxed since the amplitude of the maximum step change in the integrator input is smaller than with a 1-bit quantizer. Therefore, the integrator seldom slew limits, and the time available for linear settling is maximized. This, in turn, implies a savings in power. Finally, the amplitude of spectral tones is greatly diminished because the total quantization noise is lower, and the quantizer is not continuously in an overload condition.

The primary drawback of multibit quantization is the need for high linearity in the feedback DAC, as discussed in Section 3.3.6. Unfortunately, much of the power savings of multibit quantization are lost owing to the complexity of the techniques that must be used to linearize the output of this DAC.

The architecture adopted for the experimental modulator described herein is the 2-1 cascade shown in Figure 5.1 with $b = 2.5$, $\beta = 0.25$, and $\lambda = 1$. This modulator has a high overload level, low quantization noise, and low spectral tones. The architecture of Section 5.1.2 was not selected because of its relatively low overload level, the low peak *SNDR*, and the presence of strong spectral tones at low input levels. An architecture that employs multibit quantization, such as the second-order modulator discussed in this section, was not adopted because of the DAC implementation difficulties.

5.2 Signal Scaling

An important consideration in the design of a $\Sigma\Delta$ modulator is the size of the integrator output swings as the modulator input approaches full scale. This section discusses the implementation of the modulator of Figure 5.1 using scaling factors to reduce the output swing requirements of the integrators. The objective of signal scaling is to allow a large modulator input signal range without exceeding the linear output signal range of practical integrator implementations.

In a $\Sigma\Delta$ modulator wherein the resolution is limited by circuit noise, the dynamic range can be increased by maximizing the feedback reference voltages (the DAC output voltages) and the modulator's overload point relative to these reference levels. Without the use of voltage boosting circuits, the maximum levels for the feedback ref-

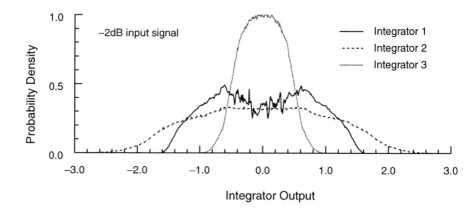

FIGURE 5.8 Probability density function of integrator outputs in the architecture of Figure 5.1 with $b = 2.5$, $\beta = 0.25$, and $\lambda = 1$.

erences are the supply voltages. In Section 5.1 it was shown that a 2-1 cascaded modulator can have an overload level of −1.3 dB. Thus, an input signal range of −1.3 dB, or 0.86, relative to the supply voltage can be achieved when using feedback reference levels equal to the supply voltages in the modulator of Figure 5.1 with $b = 2.5$, $\beta = 0.25$, and $\lambda = 1$.

Figure 5.8 shows the probability density function of the integrator outputs for the modulator of Figure 5.1 when the feedback reference levels are normalized to ±0.5. The input signal is a sinusoid with a power of −2 dB relative to an input with an amplitude of ±0.5. The output range requirement for all of the integrators is substantially greater than the feedback reference levels. However, the linear output range of an integrator circuit is typically only a fraction of the supply voltage. Thus, to implement the modulator of Figure 5.1 without saturating the integrators, and with the feedback reference levels constrained to be equal to the supply voltages, signal scaling must be employed.

Figure 5.9 illustrates the block diagram of a 2-1 cascaded modulator wherein a scaling factor is applied to each integrator input [5.3]. Integrators of this form are easily realized in switched-capacitor circuits. The system of Figure 5.9 has four more degrees of freedom than that of Figure 5.1. Therefore, after meeting the constraints imposed by the choices of b, β, and λ, the remaining degrees of freedom can be used to scale the signal levels throughout the modulator.

Signal Scaling

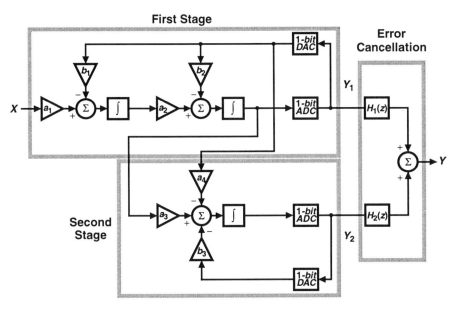

FIGURE 5.9 A 2-1 cascaded sigma-delta modulator with integrator gains.

The signal swings at the outputs of the integrators can be scaled through the feedback gain factors b_1, b_2, and b_3. As shown in Figure 5.8, when the input is a sinusoid that is 2 dB below "full scale", and the feedback references are ±0.5, the output swing of the first integrator is ±1.66, that of the second integrator is ±2.98, and that of the third integrator is ±1.30. If it is assumed that the maximum linear output range of the integrators is equal to a fraction, η, of the feedback reference levels, then in order to reduce the integrator swings to less than the output range of the integrators the scaling factors b_1, b_2, and b_3 must satisfy

$$b_1 < \frac{\eta}{2 \times 1.66} \tag{5.1}$$

$$b_2 < \frac{\eta b}{2 \times 2.98} \tag{5.2}$$

$$b_3 < \frac{\eta}{2 \times 1.30}. \tag{5.3}$$

It is advantageous from the standpoint of circuit noise, power dissipation and die area to make the two gains in the first integrator equal so that the sampling network can be shared between the input signal and the feedback signal. This imposes the additional constraint

$$a_1 = b_1. \tag{5.4}$$

The remaining integrator gains can be expressed in terms of the feedback gains, b_1, b_2, and b_3, and the systems gains, b, β, and λ. If it is assumed that the feedback reference levels provided by the outputs of the first-stage and second-stage DAC's are identical, then the remaining integrator gains must satisfy the following relationships:

$$a_2 = \frac{1}{b} \times \frac{b_2}{b_1} \tag{5.5}$$

$$a_3 = b\beta \times \frac{b_3}{b_2} \tag{5.6}$$

$$a_4 = \lambda\beta \times b_3. \tag{5.7}$$

After selecting the scaling factors b_1, b_2, and b_3, the remaining integrators gains can be determined using (5.4)-(5.7). A final consideration in choosing b_1, b_2, and b_3 is that all integrator gains should be ratios of small integers to facilitate the circuit implementation.

It will be shown in Chapter 6 that the integrator circuits used in this work can achieve an output swing of 0.8 times the feedback reference voltages. Table 5.1 lists integrator gains that meet the conditions set forth in (5.1)-(5.7) with $\eta = 0.8$ and $b = 2.5$, $\beta = 0.25$, and $\lambda = 1$. Figure 5.10 shows the probability density functions of the integrator outputs in the modulator of Figure 5.9 with the integrator gains of Table 5.1. It can be seen that all integrator outputs are well within the desired 80% of the feedback reference levels (±0.5).

5.3 Integrator Implementation

In this section, the issues of capacitor matching and settling time in a switched-capacitor integrator are explored. These discussions assume an integrator of the form

Integrator Implementation

TABLE 5.1 Integrator gains.

Integrator Gain	Value
a_1	1/5
b_1	1/5
a_2	1/3
b_2	1/6
a_3	3/4
a_4	1/20
b_3	1/5

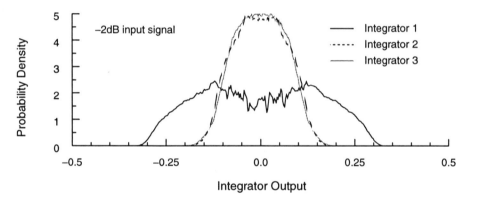

FIGURE 5.10 Probability density function of integrator outputs in the architecture of Figure 5.9 with integrator gains shown in Table 5.1.

shown in Figure 5.11. Φ1 and Φ2 are two-phase, non-overlapping clocks. When a clock signal is high, all switches controlled by it conduct, and when the clock is low the switches are open circuits. The basic operation of this circuit is as follows. When Φ1 is high, charge proportional to the input voltage is stored on the sampling capacitor, C_S. Then, when Φ2 rises, the charge on C_S is transferred to the integration capacitor, C_I.

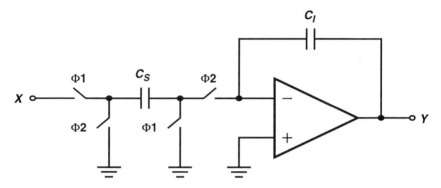

FIGURE 5.11 Simple switched-capacitor integrator.

5.3.1 Capacitor Matching

As shown in Section 3.3.5, the baseband quantization noise of a 2-1 cascaded modulator increases in the presence of mismatch between the error mixing coefficient, β, and its digital estimate, $\hat{\beta}$. In the implementation of the 2-1 cascaded modulator shown in Figure 5.9, the error mixing coefficient is realized as a function of the integrator gains as

$$\beta = \frac{b_1 a_2 a_3}{b_3}. \tag{5.8}$$

Each of the integrator gains, in turn, is realized as the ratio of capacitors

$$a_i = \frac{C_{Sai}}{C_{Ii}} \tag{5.9}$$

$$b_i = \frac{C_{Sbi}}{C_{Ii}}, \tag{5.10}$$

where C_{Sai} and C_{Sbi} are the sampling capacitors in the ith integrator for the a and b coefficients, respectively, and C_{Ii} is the integration capacitor in the ith integrator.

As will be shown in Chapter 6, the sampling capacitor for b_3 can be realized by reusing some of the unit cells that implement the sampling capacitor for a_3. Thus, the sampling capacitor for the a_3 coefficient may be expressed as

Integrator Implementation

$$C_{Sa3} = C_{Sb3} + C_{Saa3}, \tag{5.11}$$

where C_{Saa3} represents that portion of the sampling capacitor for a_3 that is not shared with the sampling capacitor for b_3. Since the integration capacitance for a_3 and b_3 is the same capacitor, it follows that the error mixing coefficient, β, may be expressed as

$$\beta = \frac{C_{Sb1}}{C_{I1}} \times \frac{C_{Sa2}}{C_{I2}} \times \left(1 + \frac{C_{Saa3}}{C_{Sb3}}\right). \tag{5.12}$$

As shown in Appendix C, if the capacitances are modeled as Gaussian random variables whose mean is much larger than their standard deviation, then β is approximately a Gaussian random variable with the fractional variance

$$\frac{\sigma[\beta]^2}{E[\beta]^2} = \frac{\sigma[C_{Sb1}]^2}{E[C_{Sb1}]^2}(1+b_1^3) + \frac{\sigma[C_{Sa2}]^2}{E[C_{Sa2}]^2}(1+a_2^3) \\ + \frac{\sigma[C_{Sa3}]^2}{E[C_{Sa3}]^2}\left\{\frac{b_3}{a_3-b_3}\left(1+\left(\frac{a_3-b_3}{b_3}\right)^3\right)\right\}, \tag{5.13}$$

where $\sigma[\cdot]^2$ and $E[\cdot]$ denote the variance and expectation operators, respectively. The variation in β is typically dominated by the variation in C_{Sa3}. This usually limits the extent to which the capacitor sizes in the third integrator can be scaled.

5.3.2 Integrator Settling

When the integrator of Figure 5.11 enters the charge transfer phase, its output approaches a final value asymptotically. Typically, the change in the output voltage is first limited by a maximum rate of change, known as the slew rate, and this slew-limited period is followed by an exponential response commonly referred to as linear settling. The exact settling characteristics of an integrator depend on the details of the integrator and amplifier implementations. Nevertheless, as shown in previous work [5.3], a good first-order model of the settling characteristics of a switched-capacitor integrator is

$$v_{out}[n+1] = v_{out}[n] + g(av_{in}[n]), \tag{5.14}$$

where

Design of a Low-Voltage, High-Resolution Sigma-Delta Modulator

$$g(x) = \begin{cases} x(1-e^{-T_S/\tau}), & |x| \leq \tau S \\ x - \text{sgn}(x)\tau S e^{\left(\frac{|x|}{\tau S} - \frac{T_S}{\tau} - 1\right)}, & \tau S \leq |x| \leq (\tau + T_S)S \\ \text{sgn}(x)ST_S, & (\tau + T_S)S < |x|. \end{cases} \quad (5.15)$$

τ is the integrator settling time constant, S is the integrator slew rate, and T_S is the sampling time. In (5.15), the integrator response is modeled with a single pole time constant and a maximum rate of change that is limited to S.

Using the single-pole integrator model of (5.14) and (5.15), functional simulations have shown [5.3] that high-resolution data conversion may be achieved in two possible settling regimes. One is a *fast regime* wherein the settling time constant, τ, is smaller than an upper limit, and the slew rate, S, is larger than a lower limit. In this regime, although slew limiting may often occur, the effects of nonlinearities introduced by slew limiting are subsequently eliminated by thorough linear settling.

The other integrator settling regime is a *slow regime* wherein both τ and S are large. Here the relationship between τ and S is such that slew limiting never occurs. This imposes a *lower* limit on τ, as well as a lower limit on S. The existence of the slow regime is due to the fact that, in the absence of slew limiting, the settling process is assumed to be entirely linear. Hence, the incomplete settling that results from a large τ manifests itself only as a reduction in integrator gain and does not introduce harmonic distortion.

In the design of the experimental modulator presented herein, an integrator that operates in the slow regime was not used for two reasons. First, nonlinearities in the nominally linear settling process would result in harmonic distortion. Second, when an amplifier whose first stage is a differential pair is used in the feedback configuration shown in Figure 4.5, the slew rate and settling time constant are related through

$$S = \left(1 + \frac{C_S + C_P}{C_I}\right) \times \frac{V_{GS} - V_T}{\tau}, \quad (5.16)$$

where V_{GS} is the gate-to-source voltage, V_T is the threshold voltage of the input transistors, and $V_{GS} - V_T$ is referred to herein as the transistor overdrive voltage. It is apparent from (5.16) that it is necessary to employ relatively large overdrive voltages to simultaneously achieve both a high slew rate and a large settling time constant. For example, for a slew rate of 60 V/µs and a settling time constant of 26 ns with C_S = 4pF, C_P = 1 pF, C_I = 20 pF, the transistor overdrive voltage must be equal to 1.25 V. Large overdrive voltages on the input transistors of amplifiers make low supply voltage implementation difficult.

5.4 Circuit Noise

The performance of a well-designed high-resolution $\Sigma\Delta$ modulator should generally be limited by thermal and flicker ($1/f$) noise in the transistors comprising the modulator [5.6]. Thermal noise is the result of the random motion of electrons in a conductor. It is proportional to absolute temperature, it has a Gaussian amplitude distribution, and its power spectral density is white up to the THz range.

In contrast to thermal noise, the source of flicker noise in MOS transistors is not well understood. It is believed to be related to imperfections in the crystal lattice at the interface between the oxide and the silicon layers [5.7], [5.8]. The power spectral density of flicker noise is approximately inversely proportional to frequency and is independent of the device biasing conditions. Furthermore, its amplitude distribution tends to vary from device to device and may not be Gaussian.

The noise of an MOS transistor may be modeled with a voltage source between the gate and source terminals or a current source between the drain and source terminals [5.6]. Since thermal and flicker noise are independent random variables, their powers add. Thus, the equivalent input-referred power spectral density of the voltage noise can be written as

$$\frac{\overline{v_{in}^2}}{\Delta f} = 4kT\gamma\frac{1}{g_m} + \frac{K_f}{WLC_{ox}f}, \qquad (5.17)$$

where k is Boltzmann's constant, T is the absolute temperature, γ is a bias and technology dependent factor [5.9], g_m is the transconductance of the device, K_f is an experimentally determined constant that is bias independent but highly technology dependent, and long-channel transistor behavior is assumed. The first term on the right hand side of (5.17) represents thermal noise, while the second term represents flicker noise. The frequency at which the thermal noise and flicker noise powers are equal is referred to as the *corner frequency* and is given by

Design of a Low-Voltage, High-Resolution Sigma-Delta Modulator

$$f_c = \frac{K_f \mu (V_{GS} - V_T)}{4kT\gamma L^2},\qquad(5.18)$$

where μ is the carrier mobility in the channel of the transistor, V_{GS} is the gate-to-source voltage, V_T is the threshold voltage, and L is the channel length.

It should be noted that, if conventional long-channel MOS device behavior is assumed, the value of γ varies from 1 when the transistor is in the linear region to 2/3 when the transistor is in the saturation region. However, experimental measurements on short-channel transistors have shown that the value of γ can increase dramatically as the device enters the saturation region [5.9]. This increase is thought to be related to hot electrons in the channel and tends to be more pronounced as the transistor drain-to-source voltage exceeds $V_{GS} - V_T$. In this work, since the supply voltage is only 1.8 V, high drain-to-source voltages do not appear across amplifier input transistors. Therefore, based on published measured data [5.9], a value of $\gamma = 1.2$ has been assumed for transistors operating in saturation.

5.4.1 Shaping of Circuit Noise

The noise shaping property of a $\Sigma\Delta$ modulator attenuates baseband circuit noise that is introduced into the forward path of the modulator after the first stage. In the architecture of Figure 5.9, the input-referred circuit noise is approximately

$$N_{in}(z) = N_1(z) + \frac{(1-z^{-1})}{a_1}N_2(z) + \frac{(1-z^{-1})^2}{a_1 a_2}N_3(z),\qquad(5.19)$$

where $N_1(z)$, $N_2(z)$, and $N_3(z)$ represent circuit noise injected at the inputs to the first, second, and third integrators, respectively. Following an approach similar to that used in Section 3.3.2, it can be shown that, when the circuit noise is white, the input-referred baseband noise power is

$$S_{Nin} = \frac{1}{M}S_{N1} + \frac{1}{a_1^2} \times \frac{\pi^2}{3M^3}S_{N2} + \frac{1}{a_1^2 a_2^2} \times \frac{\pi^4}{5M^5}S_{N3},\qquad(5.20)$$

where S_{N1}, S_{N2}, and S_{N3} are the circuit noise powers at the inputs to the first, second, and third integrators, respectively. For example, when $M = 80$, $a_1 = 1/5$, and $a_2 = 1/3$, the baseband noise introduced at the inputs to the second and third inte-

Circuit Noise

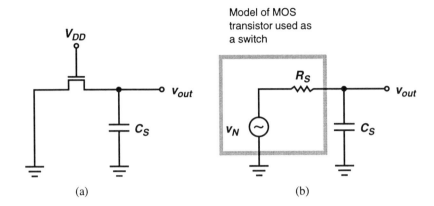

FIGURE 5.12 (a) A switched-capacitor sampling network, (b) circuit model for sampling noise.

grators is attenuated by 1.3×10^{-2} and 1.1×10^{-4}, respectively, relative to noise introduced at the input of the first integrator.

5.4.2 Sampling Noise

In this section, the effect of sampling noise is quantified for two switched-capacitor integrator circuits. First, noise introduced during the sampling phase, $\Phi 1$, and charge transfer phase, $\Phi 2$, is referred to the input of the integrator of Figure 5.11. Then, the analysis is repeated for an integrator in which the input is sampled onto an array of sampling capacitors. This latter topology is representative of the design used in the second and third integrator stages of the experimental modulator described in Chapter 6.

The circuit in Figure 5.12(a) is a simplified representation of the input sampling network in the integrator of Figure 5.11 when $\Phi 1$ is high. The power of the sampled noise can be estimated using the circuit in Figure 5.12(b), where the MOS switch is modeled as a resistor in series with a noise voltage source. If the sampling time is much longer than the time constant formed by R_S and C_S, then the voltage drop across the MOS switch is approximately zero at the end of the sampling phase. Therefore, the transistor is in the linear region of operation, and the noise factor, γ, is equal to 1. Furthermore, the switching process resets trapping states at the silicon-oxide interface, thereby preventing the accumulation of low frequency flicker noise. Thus, flicker noise can be neglected in such circuits.

To estimate the noise power at v_{out} in Figure 5.12(b), first note that the thermal noise of the resistor is white and has a one-sided power spectral density of

$$\frac{\overline{v_N^2}}{\Delta f} = 4kTR_S. \tag{5.21}$$

The broadband resistor noise in Figure 5.12(b) is filtered by the single-pole, lowpass filter formed by R_S and C_S. The total noise power at v_{out} is thus

$$\overline{v_{out}^2} = \frac{1}{2\pi} \int_0^\infty |H(j\omega)|^2 \, \overline{v_N^2} \, d\omega, \tag{5.22}$$

where

$$H(j\omega) = \frac{1}{1 + j\omega R_S C_S}, \tag{5.23}$$

and

$$|H(j\omega)|^2 = \frac{1}{1 + (\omega R_S C_S)^2}. \tag{5.24}$$

Substitution of (5.21) and (5.24) into (5.22) yields

$$\begin{aligned}
\overline{v_{out}^2} &= \frac{4kTR_S}{2\pi} \int_0^\infty \frac{1}{1 + (\omega R_S C_S)^2} d\omega \\
&= \frac{2kTR_S}{\pi} \left(\frac{1}{R_S C_S}\right) \arctan(\omega R_S C_S) \Big|_{\omega = 0}^{\omega = \infty} \\
&= \frac{kT}{C_S}.
\end{aligned} \tag{5.25}$$

Circuit Noise

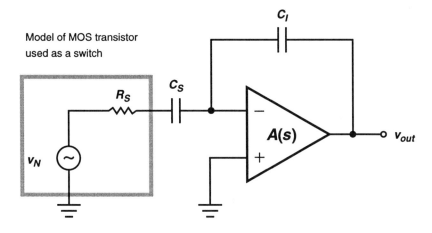

FIGURE 5.13 Sampling noise during charge transfer in a switched-capacitor integrator.

It should be noted that the value of the noise generating element, R_S, does not appear in the final expression for the output noise power in (5.25) since R_S also bandlimits the noise.

When the sampling switch in Figure 5.12(a) is opened, the noise voltage is stored on C_S. If the sampling period is much longer than the time constant formed by R_S and C_S, the high frequency components of the noise are aliased into the frequency band from 0 to f_S. As a result, the full power of the sampling noise appears in this band with an approximately white spectral density. kT/C_S is thus the total power of the input-referred sampling noise introduced during $\Phi 1$ in the integrator of Figure 5.11. Since the sampling noise is distributed uniformly across the Nyquist band, if the signal is oversampled by a factor M the thermal noise power appearing in the baseband is

$$\overline{v_M^2} = \frac{kT}{MC_S}. \tag{5.26}$$

As charge that was sampled during $\Phi 1$ is transferred from C_S to C_I during $\Phi 2$, additional thermal noise is introduced by the switches. The thermal noise introduced by the switches in the circuit of Figure 5.11 during $\Phi 2$ is bandlimited by the response of the operational amplifier as well as the time constant formed by the on-resistance of the switches and the sampling capacitor. The schematic shown in Figure 5.13 can be used to examine the impact of sampling noise during the charge transfer phase. If the single-pole amplifier model of Section 4.3 is used and if the two poles in the circuit of Figure 5.13 are widely separated, then the transfer function from v_N to v_{out} is approximately

$$H_N(s) \approx -\frac{C_S}{C_I} \times \frac{1}{(1+s/p_A)(1+s/p_S)}, \quad (5.27)$$

where

$$p_A = -\frac{g_m}{C_S}, \quad (5.28)$$

$$p_S = -\frac{1}{R_S C_S}, \quad (5.29)$$

and it has been assumed that $|p_S| \gg |p_A|$.

The total noise power at v_{out} in the circuit of Figure 5.13 can be obtained in a manner similar to that used for the circuit of Figure 5.12 and shown to be

$$\overline{v_{out}^2} = \left(\frac{C_S}{C_I}\right)^2 kTR_S \left(\frac{|p_A||p_S^2|}{|p_S^2 - p_A^2|} + \frac{|p_S||p_A^2|}{|p_A^2 - p_S^2|}\right). \quad (5.30)$$

If $p_S \gg p_A$, then the input-referred noise power during $\Phi2$ is

$$\overline{v_{in}^2} \approx kTR_S \times |p_A| = \frac{kT}{C_S}\left(\frac{p_A}{p_S}\right). \quad (5.31)$$

If the sampling period is much longer than the reciprocal of $|p_A|$, then the high frequency components of the noise are aliased into the frequency band from 0 to f_S. Therefore, the sampling noise appears in this band with an approximately white spectral density, and (5.31) represents the total power of the input-referred sampling noise introduced during $\Phi2$ in the integrator of Figure 5.11.

Since the sampling noise voltages introduced during $\Phi1$ and $\Phi2$ are uncorrelated, the noise powers add. Therefore, the total input-referred sampling noise power in the integrator of Figure 5.11 is

Circuit Noise

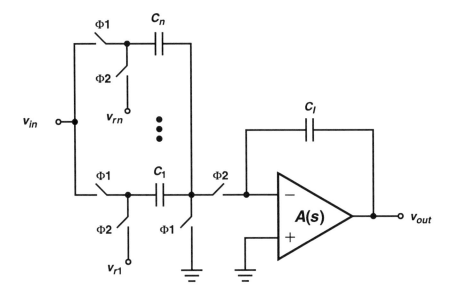

FIGURE 5.14 Sampling noise when using multiple sampling capacitors.

$$\overline{v_{in}^2} = \frac{kT}{C_S}\left(1 + \frac{p_A}{p_S}\right). \tag{5.32}$$

As described in Chapter 6, switched-capacitor integrator implementations of the form shown in Figure 5.14 are adopted for the second and third integrators of the experimental modulator. Therefore, the sampling noise of this circuit is examined here. The z-domain representation of the output of this circuit is

$$V_{out}(z) = \frac{z^{-1}}{1-z^{-1}}\sum_{j=1}^{N}\left(\frac{C_j}{C_I}\right)V_{in}(z) - \frac{z^{-1/2}}{1-z^{-1}}\sum_{j=1}^{N}\left(\frac{C_j}{C_I}V_{rj}(z)\right). \tag{5.33}$$

It will be shown here that the total input-referred sampling noise is inversely proportional to the sum of all the sampling capacitors. Thus, no sampling noise penalty is incurred by dividing a single sampling capacitor into smaller capacitors in parallel.

As shown in (5.25), during Φ1 each capacitor samples a noise of

Design of a Low-Voltage, High-Resolution Sigma Delta Modulator

Design of a Low-Voltage, High-Resolution Sigma-Delta Modulator

$$\overline{v_{1j}^2} = \frac{kT}{C_j}. \tag{5.34}$$

As in the circuit of Figure 5.11, the operational amplifier in the integrator of Figure 5.14 limits the noise bandwidth during $\Phi 2$. Thus, each of the input capacitors samples a noise power of

$$\overline{v_{2j}^2} = \frac{kT}{C_j}\left(\frac{p_A}{p_S}\right), \tag{5.35}$$

where

$$p_A = -\frac{g_m}{\sum_{j=1}^{N} C_j}, \tag{5.36}$$

and

$$p_S = -\frac{1}{R_j C_j} \tag{5.37}$$

and is assumed to be a constant for all j.

If $V_{n1j}[n]$ and $V_{n2j}[n]$ are sequences of uncorrelated, zero-mean random variables whose variances are given by (5.34) and (5.35), respectively, then the z-domain output of the integrator of Figure 5.14 with v_{in} and v_{rj} set to zero is given by

$$V_{out}(z) = \frac{z^{-1}}{1-z^{-1}} \sum_{j=1}^{N} \left(\frac{C_j}{C_I} V_{n1j}(z)\right) - \frac{z^{-1/2}}{1-z^{-1}} \sum_{j=1}^{N} \left(\frac{C_j}{C_I} V_{n2j}(z)\right), \tag{5.38}$$

Referring the noise to the input of the integrator results in

$$V_{N,in}(z) \approx \frac{\sum_{j=1}^{N} C_j V_{n1j}(z)}{\sum_{j=1}^{N} C_j} - \frac{\sum_{j=1}^{N} C_j V_{n2j}(z)}{\sum_{j=1}^{N} C_j}. \tag{5.39}$$

Circuit Noise

The power in the input-referred noise power can now be expressed [5.10] as a function of the variances of $V_{n1j}[n]$ and $V_{n2j}[n]$ to yield

$$\overline{v_{N,in}^2} = \frac{\sum_{j=1}^{N} C_j^2\left(\frac{kT}{C_j}\right)}{\left(\sum_{j=1}^{N} C_j\right)^2} + \frac{\sum_{j=1}^{N} C_j^2\left(\frac{kT}{C_j}\left(\frac{p_A}{p_S}\right)\right)}{\left(\sum_{j=1}^{N} C_j\right)^2}. \tag{5.40}$$

Simplifying (5.40) results in

$$\overline{v_{N,in}^2} = \frac{kT}{\sum_{j=1}^{N} C_j}\left(1 + \frac{p_A}{p_S}\right). \tag{5.41}$$

Thus, the input-referred sampling noise in the circuit of Figure 5.14 is consistent with sampling the input onto a single capacitor with a capacitance equal to the sum of all the sampling capacitors.

Computing the kT/C noise in more elaborate switched-capacitor networks can be considerably more complicated than the above analyses because integrating the magnitude squared of the noise transfer function can be difficult. In this text, the results derived in this section are used to estimate the sampling noise in the circuits that comprise the modulator.

5.4.3 Amplifier Thermal Noise

The thermal noise in the transistors used to implement an amplifier may be represented by an equivalent noise source the amplifier's input. When the thermal noise of the operational amplifier embedded in the integrator in Figure 5.15 is dominated by the noise of an input differential-pair with a current-mirror load, then integrating this noise over all frequency and referring it to an equivalent noise source at the amplifier's noninverting input terminal results in [5.6]

$$\overline{v_N^2} = \frac{2kT}{C_C} \times \frac{\gamma}{1 + C_S/C_I} \times \left(1 + \frac{g_{m3}}{g_{m1}}\right), \tag{5.42}$$

Design of a Low-Voltage, High-Resolution Sigma-Delta Modulator

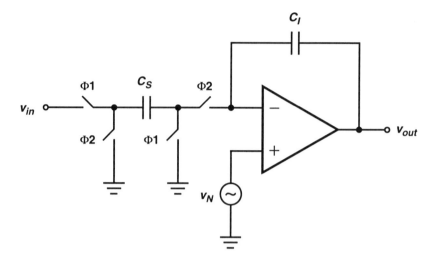

FIGURE 5.15 Amplifier noise during the charge transfer phase.

where C_C is the amplifier's compensation capacitor, γ is the short-channel noise enhancement factor, C_S is the capacitor to ground at the inverting input terminal of the amplifier, C_I is the integrating feedback capacitor, g_{m1} is the transconductance of the input transistors, and g_{m3} is the transconductance of the load devices.

The output of the integrator of Figure 5.15 may be expressed as a function of the input signal and the amplifier noise as

$$V_{out}(z) = \frac{z^{-1}}{1-z^{-1}} \times \frac{C_S}{C_I} V_{in}(z) + \left(1 + \frac{z^{-1/2}}{1-z^{-1}} \times \frac{C_S}{C_I}\right) V_N(z). \tag{5.43}$$

Referring the thermal noise of the amplifier to the integrator's input results in

$$V_{N,in}(z) \approx \left[1 + \frac{C_I}{C_S}(1-z^{-1})\right] V_N(z). \tag{5.44}$$

From this relationship it is apparent that when the thermal noise of the amplifier is referred to the input of a switched-capacitor integrator with a large oversampling

Circuit Noise

ratio, that noise is insensitive to the integrator gain. Thus, the use of a small gain in the first integrator does not significantly increase the input-referred thermal noise.

5.4.4 Flicker Noise

In addition to thermal noise, flicker noise in the constituent MOS transistors can result in a significant flicker noise component when referred to the amplifier input. As shown in (5.17), the flicker noise of an MOS transistor is inversely proportional to its gate area. Thus, the $1/f$ noise in an amplifier can be reduced simply by increasing the size of transistors that contribute significantly to the input-referred flicker noise. If the flicker noise factor, K_f, of a process is large, the use of transistor sizing to suppress flicker noise may require overly large input transistors. As discussed in Section 4.3, large input transistors can degrade the settling performance by increasing the amplifier's input capacitance.

Another approach to attenuating flicker noise is chopper stabilization [5.11]. In this method, the input signal is modulated to a chopping frequency above the noise corner frequency of the amplifier before the amplifier flicker noise is added to it. Subsequently, the integrator output is demodulated, restoring the input signal back to its original frequency while translating the flicker noise to the chopping frequency. The flicker noise can then be removed with a low pass filter. Unfortunately, chopper stabilization typically forces the amplifier's outputs to undergo large swings from one cycle to the next and, therefore, imposes severe settling constraints.

Correlated double sampling (CDS) is another technique for low frequency noise reduction in oversampling systems [5.12]. CDS is implemented by storing the integrator noise voltage in one clock cycle, then subtracting it from the stored value on the following cycle, thus introducing a zero at dc in the noise transfer function. Since the flicker noise power is concentrated at low frequencies, CDS reduces this noise to a negligible level. Unfortunately, the bottom-plate capacitance associated with the storage capacitor increases the integrator input capacitance, thereby degrading the settling response. Moreover, the additional switches needed in a CDS integrator increase the kT/C noise. Finally, the thermal noise of an amplifier referred to the input of a CDS integrator is a function of the correlation frequency and can be higher than that noise in an integrator that does not employ CDS [5.4].

In this work, two noise reduction methods were employed in the first integrator: transistor sizing and CDS. It was found that transistor sizing alone provided adequate suppression of flicker noise, and the circuits that relied on this technique offered the best performance.

5.5 Modulator Specifications

The performance targets for the experimental $\Sigma\Delta$ modulator described herein were outlined at the beginning of this chapter. In this section, specifications for key parameters of the circuits comprising the modulator are established. These include requirements for capacitor matching, integrator settling, amplifier gain, kT/C noise, thermal noise, and flicker noise.

5.5.1 Matching Requirements

In Section 3.3.5 it was shown that a matching error of 6.8% between the coefficient β and its digital estimate $\hat{\beta}$ would yield a 1-dB reduction in the dynamic range of the proposed $\Sigma\Delta$ modulator. The relationship between the fractional variance in β and sampling capacitors in the integrators was derived in Section 5.3. In this section, a lower bound on integrator capacitor sizes is obtained by using a 6.8% target for variations in β and making some assumptions about the matching characteristics of capacitors. If (5.13) is used with the values shown in Table 5.1 for the gain factors, the fractional variance becomes

$$\frac{\sigma[\beta]^2}{E[\beta]^2} = 1.008\frac{\sigma[C_{Sb1}]^2}{E[C_{Sb1}]^2} + 1.037\frac{\sigma[C_{Sa2}]^2}{E[C_{Sa2}]^2} + 29.72\frac{\sigma[C_{Sa3}]^2}{E[C_{Sa3}]^2}. \tag{5.45}$$

The strong dependence on the fractional variance of C_{Sa3} is a consequence of the choices for a_3 and b_3 and can be reduced by reducing the ratio a_3/b_3.

Previous work has suggested that the variance of capacitor sizes is inversely proportional to the capacitor area, and hence to the capacitance [5.13]. To estimate the size of the capacitors required on the basis of capacitor matching, it is assumed that the 3-σ fractional standard deviation for 0.5-pF poly-to-metal capacitors in a 0.8-μm CMOS process is at worst 1% [5.14]. It then follows from (5.45) that to limit the fractional standard deviation in β to less than 6.8% due to the variation in each of the sampling capacitors, the following limits on the capacitor size must be observed

$$C_{S1} > 11 \text{ fF} \tag{5.46}$$

$$C_{S2} > 11 \text{ fF} \tag{5.47}$$

$$C_{S3} > 321 \text{ fF}, \tag{5.48}$$

where C_{S1}, C_{S2}, and C_{S3} are the sampling capacitors of the first, second, and third integrators, respectively. As will be shown in Section 5.5.4, C_{S1} and C_{S2} must actually be much larger than indicated in (5.46) and (5.47) to meet the sampling noise requirements. Therefore, in practice the variation in β is dominated by the variation in C_{S3}.

5.5.2 Settling Time

The maximum settling time constant, τ, and minimum slew rate, S, are determined in this section for each integrator in the modulator of Figure 5.9 when the gains listed in Table 5.1 are used. These integrator settling requirements were estimated using functional simulations that model the settling behavior of the integrator using (5.15). τ and S are assumed to be related through (5.16), and the transistor overdrive, $V_{GS} - V_T$, has been chosen to be 150 mV. A smaller overdrive voltage would require larger input transistors, thereby increasing the integrator's input capacitance, while a larger overdrive is difficult to accommodate within a 1.8-V power supply. The requirements identified in this section ensure that the performance of the modulator does not degrade significantly beyond what could be achieved with integrators that settle instantaneously.

For the first integrator, if it is assumed that $C_S = 4$ pF and $C_P = 1$ pF, simulation results indicate that with

$$S > 12.5 \text{ V/}\mu\text{s} \tag{5.49}$$

$$\tau < 15 \text{ ns}, \tag{5.50}$$

there is negligible degradation in the dynamic range of the modulator.

Likewise, if it is assumed that $C_S = 1.6$ pF and $C_P = 0.7$ pF, simulation results suggest that in the second integrator with

$$S > 6.3 \text{ V/}\mu\text{s} \tag{5.51}$$

$$\tau < 35 \text{ ns}, \tag{5.52}$$

the modulator's dynamic range does not deteriorate.

Finally, simulation results for the third integrator with $C_S = 1.5$ pF and $C_P = 0.3$ pF indicate that with

$$S > 4.1 \text{ V/µs} \tag{5.53}$$

$$\tau < 70 \text{ ns}, \tag{5.54}$$

there is negligible degradation in the dynamic range of the modulator.

5.5.3 Amplifier Gain Requirements

As discussed in Section 3.3.5, integrator leak, which is a consequence of finite amplifier dc gain, limits the extent to which low-frequency quantization noise is shaped in a $\Sigma\Delta$ modulator, thereby increasing the baseband noise. The transfer function of a leaky integrator is

$$H(z) = \frac{z^{-1}}{1-(1-\varepsilon)z^{-1}}, \tag{5.55}$$

where ε is referred to as the leakage factor.

In a switched-capacitor integrator of the form shown in Figure 5.16 the integrator leakage factor is

$$\varepsilon \approx \frac{1}{A_0+1} \times \frac{C_S}{C_I}, \tag{5.56}$$

where A_0 is the amplifier's open loop gain. The analysis in Section 3.3.5 revealed that the leakage factor must be less than 0.002 in the first and second integrators and less than 0.017 in the third integrator. Therefore, for the integrator gains shown in Table 5.1, the minimum amplifier gains are 100, 167, and 44 for the first, second, and third amplifiers, respectively.

Practical implementations of switched-capacitor circuits require amplifier gains much larger than those dictated by integrator leak in order to insure sufficient linearity and parasitic insensitivity in the integrator response. In previous work, an amplifier gain of 60 dB has proven adequate in high-resolution applications [5.15]. Therefore, the specification for this design is an open-loop gain in excess of 60 dB for all three amplifiers.

Modulator Specifications

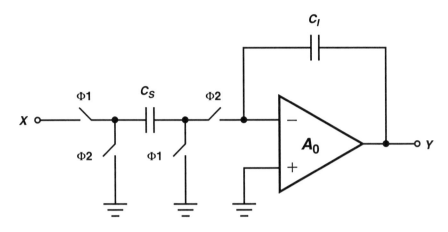

FIGURE 5.16 Simple switched-capacitor integrator with finite amplifier gain.

5.5.4 Sampling Noise Budget

A noise floor of −100dB is the objective for the experimental modulator designed in this work. Since the quantization noise floor of the 2-1 cascaded modulator has been estimated through simulations to be −105dB, the sum of the sampling noise and amplifier noise must be less than −101.7dB. If it is assumed that the sampling and amplifier noise powers are equal, then each of them must be less than −104.7dB to meet the overall noise floor requirements.

According to (5.21) and (5.31) the sampling and integration operations in the single-ended integrator of Figure 5.16 introduce an input referred thermal noise of

$$\overline{v_{NS}^2} = \frac{kT}{C_S}\left(1 + \frac{p_A}{p_S}\right), \quad (5.57)$$

where p_A and p_S have been defined in (5.28) and (5.29), respectively. When $p_A/p_S = 0.2$ and fully differential circuits are used, the sampling noise in the switched-capacitor integrator becomes

$$\overline{v_{NS}^2} = \frac{2.4kT}{C_S}, \quad (5.58)$$

where the noise has been increased by a factor of two to account for the additional noise present in a differential configuration.

To evaluate the impact of sampling noise from each of the three integrators on the baseband modulator output, the relationship given in (5.20) and the integrator gains of Table 5.1 are used. To limit the baseband noise contribution of the subsequent integrator stages to a small fraction of the first integrator, the input-referred noise budget for the second integrator is 1/10 and for the third integrator 1/100 the noise in the first integrator. With these noise targets, the sampling capacitor size constraints are

$$C_{S1} > 2.4 \text{ pF} \tag{5.59}$$

$$C_{S2} > 312 \text{ fF} \tag{5.60}$$

$$C_{S3} > 26 \text{ fF}, \tag{5.61}$$

where C_{S1}, C_{S2}, and C_{S3} are the sampling capacitors of the first, second, and third integrators, respectively.

5.5.5 Thermal Noise Budget

As discussed in Section 5.4.3, the thermal noise in the first integrator of a $\Sigma\Delta$ modulator, when referred to the modulator's input, is given approximately by (5.42) when the oversampling ratio is large. The baseband integrator thermal noise can be made approximately equal to the sampling noise by setting the compensation capacitor equal to the sampling capacitor ($C_C = C_S$). This approach was adopted in the first two integrators of the experimental prototype. Since thermal noise in the third integrator is of little consequence, C_C was scaled more aggressively in this circuit.

5.5.6 Flicker Noise Attenuation

To determine the input-referred baseband integrator flicker noise power, the power spectral density of the noise must be integrated over a particular frequency band, from f_L to f_H. The lower limit of integration, f_L, depends on the lowest frequency of interest in the desired application. Since the flicker noise power falls with frequency, the aliasing of flicker noise at multiples of the sampling frequency has a negligible impact in highly oversampled systems; thus, the upper limit, f_H, is the Nyquist frequency, f_N. The flicker noise power in a fully-differential, two-stage amplifier wherein the noise

Summary

is dominated by that of a PMOS differential input pair with NMOS current source loads is

$$\overline{v_{Nf}^2} = \frac{2K_{fP}}{W_1 L_1 C_{ox}} \left[1 + \frac{K_{fN} \mu_N L_1^2}{K_{fP} \mu_P L_3^2} \right] \ln\left(\frac{f_N}{f_L}\right), \tag{5.62}$$

where K_{fN} and K_{fP} are the flicker noise coefficients of the NMOS and PMOS transistors, respectively, and μ_N and μ_P are the carrier mobilities of the devices. W_1 and L_1 are the channel width and length of the input PMOS devices, and L_3 is the channel length of the load devices. In digital-audio applications, the lowest frequency of interest is 20 Hz.

Transistors in the 0.8-μm CMOS technology used to implement the experimental modulator have very low flicker noise coefficients. Thus, the transistor sizes used in the amplifier should be sufficient to suppress flicker noise to a level well below the thermal noise. As a precaution, a CDS version of the first integrator was also designed. However, measured results demonstrate that CDS was not needed to attenuate the amplifier flicker noise.

5.6 Summary

In this chapter, functional simulations have been used to assess the suitability of several ΣΔ modulator architectures for digital-audio data acquisition with an oversampling ratio of 80. A 2-1 cascaded modulator has been adopted due to its ease of implementation, its overload characteristics, and the absence of spurious tones. Signal scaling has been used in the 2-1 cascade to constrain the internal signal swings, making it possible to maximize the full-scale input range. Implementation requirements including capacitor matching, integrator settling, amplifier gain, and circuit noise have been analyzed, and modulator specifications that rely on these analyses have been proposed. The next chapter describes the implementation of an experimental digital-audio band ΣΔ modulator based on these specifications.

REFERENCES

[5.1] J. Candy and G. Temes, "Oversampling methods for A/D and D/A conversion," in *Oversampling Delta-Sigma Data Converters*, pp. 1-29, New York: IEEE Press, 1992.

[5.2] K. Chao, S. Nadeem, W. Lee, and C. Sodini, "A higher order topology for interpolative modulators for oversampling A/D converters," *IEEE Trans. on Circuits and Systems II*, vol. CAS-37, pp. 309–318, March 1990.

[5.3] L. Williams and B. A. Wooley, "A third-order sigma-delta modulator with extended dynamic range," *IEEE J. Solid-State Circuits*, vol. SC-29, pp. 193-202, March 1994.

[5.4] L. Williams, *Modeling and Design of High-Resolution Sigma-Delta Modulators*, Ph.D. Thesis, Stanford University, Technical Report No. ICL93-022, August 1993.

[5.5] S. Ardalan and J. Paulos, "An analysis of nonlinear behavior in delta-sigma modulators," *IEEE Trans. on Circuits and Systems II*, vol. CAS-34, pp. 593–603, June 1987.

[5.6] P. Gray and R. Meyer, *Analysis and Design of Analog Integrated Circuits*, John Wiley & Sons, 1993.

[5.7] M. Das and J. Moore, "Measurements and interpretation of low-frequency noise in FET's," *IEEE Trans. on Electron Devices*, vol. ED-21, no. 4, pp. 247-257, April 1974.

[5.8] J. Chang, A. Abidi, and C. Viswanathan, "Flicker noise in CMOS transistors from subthreshold to strong inversion at various temperatures," *IEEE Trans. on Electron Devices*, vol. ED-41, no. 11, pp. 1965-1971, November 1994.

[5.9] A. Abidi, "High frequency noise measurements on FET's with small dimensions," *IEEE Trans. on Electron Devices*, vol. ED-33, pp. 1801-1805, November 1986.

[5.10] A. Leon-Garcia, *Probability and Random Processes for Electrical Engineers*, Addison-Wesley, 1994.

[5.11] K. Hsieh, P. Gray, D. Senderowicz, and D. Messerschmitt, "A low-noise chopper-stabilized differential switched-capacitor filtering technique," *IEEE J. Solid-State Circuits*, vol. SC-16, pp. 708–715, December 1981.

[5.12] K. Nagaraj, J. Vlach, T. R. Viswanathan, and K. Singhal, "Switched-capacitor integrator with reduced sensitivity to amplifier gain," *Electronic Letters*, vol. 22, pp. 1103-1105, October 1986.

[5.13] M. Pelgrom, A. Duinmaijer, and A. Welbers, "Matching properties of MOS transistors," *IEEE J. Solid-State Circuits*, vol. SC-24, pp. 1433-9, October 1989.

[5.14] M. Pelgrom, Philips Research Labs, Eindhoven, The Netherlands, private communication, 1996.

[5.15] B. Boser and B. Wooley, "The design of sigma-delta modulation analog-to-digital converters," *IEEE J. Solid-State Circuits*, vol. SC-23, December 1988.

CHAPTER 6
Implementation of an Experimental Low-Voltage Modulator

An experimental implementation of the ΣΔ modulator architecture described in Chapter 5 has been integrated in a 0.8-μm CMOS process with poly-to-metal capacitors and operates from a nominal supply voltage of 1.8 V. While the entire analog signal path and the clock circuits for this modulator have been integrated on a single chip, much of the digital postprocessing, including the error cancellation networks and the digital decimation filter, has been implemented in software. The setup used to test the experimental prototype is described in Appendix D.

Fully differential analog circuits are used throughout the experimental modulator for two reasons. First, a fully differential topology provides for high rejection of common-mode disturbances such as supply and substrate noise, and it is relatively immune to switch charge injection errors. Second, for a given voltage range a differential signal has twice the amplitude and four times the power of a single-ended signal, while introducing only twice the noise power since the signal flows along two independent paths. Thus, there is a net increase in dynamic range of 3 dB in fully differential circuits.

Throughout this chapter, the "positive" and "negative" components of a differential signal are denoted by the use of the subscripts "p" and "n", respectively. For example, the signal V_i has two components, V_{ip} and V_{in}, where $V_i = V_{ip} - V_{in}$.

Implementation of an Experimental Low-Voltage Modulator

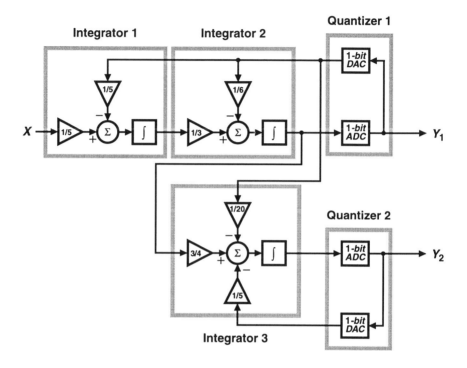

FIGURE 6.1 Block diagram of the experimental sigma-delta modulator.

A block diagram of the analog core of the experimental cascaded $\Sigma\Delta$ modulator is shown in Figure 6.1. The circuit blocks implemented in the experimental circuit but not shown in this figure are the clock generators, clock drivers and the output buffers.

This chapter describes the design of the circuits used in the experimental modulator. First, the integrator circuits are presented, and their constituent amplifiers are described. The quantizers are discussed next, followed by a description of the clock generators, the clock voltage boosting circuits and the output buffers. A brief discussion is then devoted to the impact of using an oversampling ratio that is not power of 2 on the implementation of digital decimation filters. The measured results of the experimental prototype are then presented, and finally, a figure of merit is proposed to assess the power efficiency of various A/D converters.

The Integrators

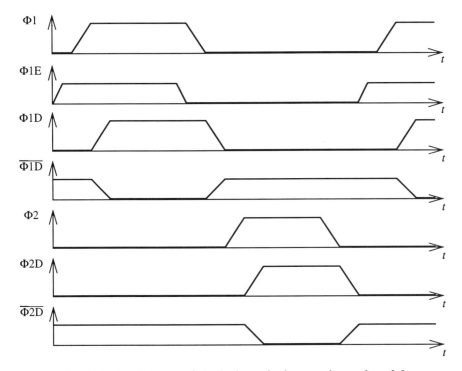

FIGURE 6.2 Timing diagram of clock phases in the experimental modulator.

6.1 The Integrators

The performance of practical $\Sigma\Delta$ modulators is generally limited by their constituent integrator circuits. To ensure adequate integrator performance, generous safety margins have been adopted in the design of the experimental modulator relative to the minimum circuit requirements identified in Section 5.5.

The three integrators and the two quantizers used in the 2-1 cascaded modulator of Figure 6.1 are controlled by two-phase, non-overlapping clocks, $\Phi 1$ and $\Phi 2$, an early clock phase, $\Phi 1E$, delayed clock phases, $\Phi 1D$ and $\Phi 2D$, and complements of the delayed phases, $\overline{\Phi 1D}$ and $\overline{\Phi 2D}$. A timing diagram illustrating these clock signals is shown in Figure 6.2. Voltage doublers drive $\Phi 1$, $\Phi 2$, $\Phi 1D$ and $\Phi 2D$, which control NMOS transistors used as switches, above the supply voltage, while simple inverters drive the other phases. $\overline{\Phi 1D}$ and $\overline{\Phi 2D}$ drive the PMOS transistors in CMOS switches,

Implementation of an Experimental Low-Voltage Modulator

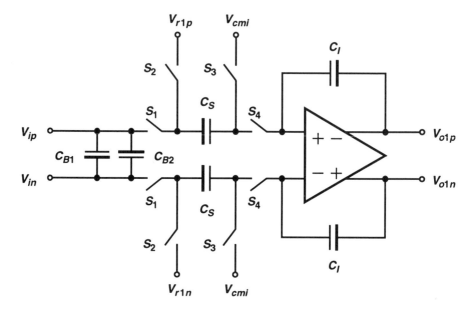

FIGURE 6.3 The first integrator without CDS.

while $\Phi 1E$ strobes the comparators. The clock generation and bootstrapping circuits are described in Sections 6.4 and 6.5, respectively.

6.1.1 The First Integrator

As noted in Section 5.4.4, two versions of the experimental modulator were implemented. One relies on sizing the transistors in the operational amplifier of the first integrator to adequately suppress flicker noise, while the other employs correlated double sampling (CDS) in this integrator. A diagram of the circuit for the former implementation is shown in Figure 6.3, and the implementation of the switches in this circuit is summarized in Table 6.1. Experimental results have shown that transistor sizing alone provides adequate low-frequency noise suppression, and that the integrator of Figure 6.3 results in a thermal noise floor for the modulator that is 2 dB lower than that obtained with the CDS integrator.

The integrator of Figure 6.3, operates as follows. During $\Phi 1$, the input voltage, V_i, is sampled onto C_S. During $\Phi 2$, charge proportional to the difference between V_i and the

The Integrators

TABLE 6.1 Switches in the first integrator.

Switch	Clock	Type	Size
S_1	$\Phi 1D$ / $\overline{\Phi 1D}$	CMOS	NMOS: 16μm/0.8μm PMOS: 40μm/0.8μm
S_2	$\Phi 2D$ / $\overline{\Phi 2D}$	CMOS	NMOS: 16μm/0.8μm PMOS: 40μm/0.8μm
S_3	$\Phi 1$	NMOS	NMOS: 32μm/0.8μm
S_4	$\Phi 2$	NMOS	NMOS: 32μm/0.8μm

TABLE 6.2 Capacitor sizes in the first integrator.

Capacitor	Size
C_S	4 pF
C_I	20 pF
C_{B1}, C_{B2}	1 pF

feedback reference voltage, V_{r1}, is transferred from C_S to C_I. Switches S_1 and S_2 are implemented as CMOS transmission gates in order to ensure a small variation in on-resistance across the full input signal range. This serves to reduce signal-dependent charge injection from switches S_3 and S_4 onto C_S and C_I to a negligible level [6.1].

In Section 5.5.4, the minimum sampling capacitance needed to meet the stated performance objectives was found to be 2.4 pF, based on kT/C noise considerations. To provide a margin of safety, a sampling capacitor of 4 pF is used. To achieve an integrator gain of 1/5, the integration capacitor, C_I, is then 20 pF. Capacitors C_{B1} and C_{B2} ensure that disturbances through the package lead frame, pins, and bond wires onto one of the two differential input lines appear on the other line as well and hence become a common-mode input to the modulator. The capacitor values for the first integrator are summarized in Table 6.2.

To enable operation of the amplifier's input stage from a 1.8-V supply, an input common-mode voltage, V_{cmi}, of only 400 mV is used. This low input common-mode voltage also permits switches S_3 and S_4 to be implemented using only NMOS transistors. The voltage used for V_{cmi} is bounded from below by concern that the source/drain junctions of S_3 and S_4 might forward bias during switching transients, which would lead to charge loss and distortion.

Implementation of an Experimental Low-Voltage Modulator

Implementation of an Experimental Low-Voltage Modulator

When both the input and feedback signals are sampled onto the same sampling capacitor, C_S, as in Figure 6.3, less kT/C noise is generated than when separate sampling capacitors are used for each signal. However, sharing the sampling capacitor introduces signal dependent loading on the feedback reference voltages. In effect, the switching process creates a charge transfer path between V_i and V_{r1}. If there is insufficient time for the disturbance on V_{r1} to settle, distortion and mixing of high frequency noise into the baseband may result. In this work, the impedance of the reference source was carefully taken into account in order to ensure that the reference voltage settled fully before the end of $\Phi 2$.

6.1.2 The Second Integrator

A diagram of the second integrator is shown in Figure 6.4, the implementation of the switches is summarized in Table 6.3, and the capacitor values are given in Table 6.4.

TABLE 6.3 Switches in the second integrator.

Switch	Clock	Type	Size
S_5	$\Phi 1D$, $\overline{\Phi 1D}$	CMOS	NMOS: 8μm/0.8μm PMOS: 8μm/0.8μm
S_6	$\Phi 2D$	NMOS	NMOS: 8μm/0.8μm
S_7	$\Phi 1D$, $\overline{\Phi 1D}$	CMOS	NMOS: 8μm/0.8μm PMOS: 8μm/0.8μm
S_8	$\Phi 2D$, $\overline{\Phi 2D}$	CMOS	NMOS: 8μm/0.8μm PMOS: 8μm/0.8μm
S_9	$\Phi 1$	NMOS	NMOS: 8μm/0.8μm
S_{10}	$\Phi 2$	NMOS	NMOS: 8μm/0.8μm

The z-domain output of this integrator is

$$V_{o2}(z) = \frac{C_{S1} + C_{S2}}{C_I} z^{-1} V_{o1}(z) - \frac{C_{S2}}{C_I} z^{-1/2} V_{r1}(z). \tag{6.1}$$

Since in the experimental prototype $z^{-1/2} V_{r1}(z) = z^{-1} V_{r1}(z)$, the circuit of Figure 6.4 does implement the transfer function required in the second integrator.

The Integrators

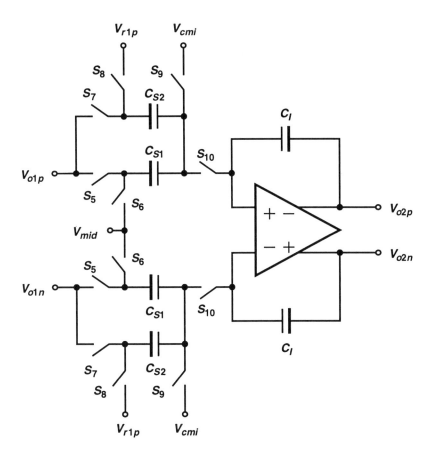

FIGURE 6.4 The second integrator.

TABLE 6.4 Capacitor sizes in the second integrator.

Capacitor	Size
C_{S1}	0.8 pF
C_{S2}	0.8 pF
C_I	4.8 pF

Implementation of an Experimental Low-Voltage Modulator

Implementation of an Experimental Low-Voltage Modulator

In Figure 6.4, the input common-mode voltage, V_{cmi}, is 400 mV, and the mid-supply reference voltage, V_{mid}, is 900 mV. This circuit may be simplified slightly by replacing the two S_6 switches that connect the bottom plates of capacitors C_{S1} to V_{mid} with a single S_6 switch that shorts the bottom plates together.

As shown in Section 5.5.4, the sampling capacitor for the second integrator must be larger than 312 fF because of kT/C noise considerations. In the integrator of Figure 6.4 the sampling capacitor is composed of the parallel combination of C_{S1} and C_{S2}, which totals 1.6 pF.

6.1.3 The Third Integrator

A diagram of the third integrator is shown in Figure 6.5. The implementation of the switches is described in Table 6.5, and the capacitor values are given in Table 6.6. The z-domain output of this integrator is

$$V_{o3}(z) = \frac{C_{S1} + C_{S2} + C_{S3}}{C_I} z^{-1} V_{o2}(z) - \frac{C_{S3}}{C_I} z^{-1/2} V_{r1}(z) - \frac{C_{S2}}{C_I} z^{-1/2} V_{r2}(z). \quad (6.2)$$

Since $z^{-1/2} V_{r1}(z) = z^{-1} V_{r1}(z)$ and $z^{-1/2} V_{r2}(z) = z^{-1} V_{r2}(z)$ in the experimental modulator, the circuit of Figure 6.5 does implement the transfer function specified for the third integrator.

As in the second integrator, the input common-mode voltage, V_{cmi}, is 400 mV, and the mid-supply reference voltage, V_{mid}, is 900 mV. The circuit of Figure 6.5 may also be simplified slightly by replacing the two S_{12} switches that connect the bottom plates of capacitors C_{S1} to V_{mid} with a single switch that shorts the bottom plates together.

The size of the sampling capacitors in the third integrator is constrained by matching considerations to be at least 321 fF, as was shown in Section 5.5.1. In the integrator of Figure 6.5 the sampling capacitor is composed of the parallel combination of C_{S1}, C_{S2} and C_{S3}, which results in a total capacitance of 1.5 pF.

The Integrators

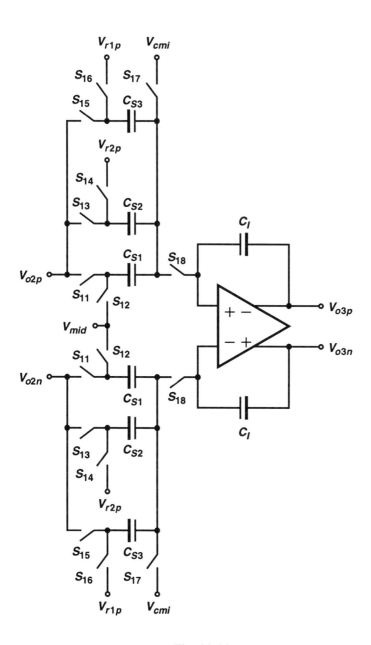

FIGURE 6.5 The third integrator.

Implementation of an Experimental Low-Voltage Modulator   **123**

TABLE 6.5 Switches in the third integrator.

Switch	Clock	Type	Size
S_{11}	Φ1D	NMOS	NMOS: 4μm/0.8μm
S_{12}	Φ2D	NMOS	NMOS: 4μm/0.8μm
S_{13}	Φ1D	NMOS	NMOS: 4μm/0.8μm
S_{14}	Φ2D	NMOS	NMOS: 4μm/0.8μm
S_{15}	Φ1D	NMOS	NMOS: 4μm/0.8μm
S_{16}	Φ2D	NMOS	NMOS: 4μm/0.8μm
S_{17}	Φ1	NMOS	NMOS: 4μm/0.8μm
S_{18}	Φ2	NMOS	NMOS: 4μm/0.8μm

TABLE 6.6 Capacitor sizes in the third integrator.

Capacitor	Size
C_{S1}	1 pF
C_{S2}	0.4 pF
C_{S3}	0.1 pF
C_I	2 pF

6.2 The Operational Amplifiers

The integrators in the experimental modulator have each been implemented using the two-stage, class A/AB operational amplifier topology shown in Figure 6.6. The basic operation of this circuit is described in Section 4.4, and key performance parameters are summarized in Table 4.1. In this section, the behavior of the amplifier is considered in greater detail.

The use of PMOS input transistors in the amplifier of Figure 6.6 makes it possible to use a common-mode input level, V_{cmi}, close to ground. This, in turn, allows for the use of relatively small NMOS transistors in implementing the switches that are connected to V_{cmi} in the integrator circuit. The sources of the input transistors, M_1 and M_2, should be tied to the n-well for improved power supply rejection [6.2]. The compensation network for the amplifier in Figure 6.6 comprises transistor M_z and capaci-

The Operational Amplifiers

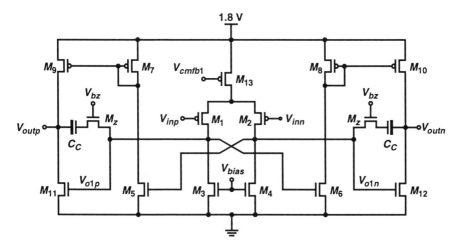

FIGURE 6.6 A two-stage class A/AB operational amplifier.

tor C_C. M_z operates in the triode region and cancels the right half plane zero. Owing to the low supply voltage, the gate of M_z is driven from the on-chip, low-ripple voltage doubler described in Section 6.5.1. A possible alternative to this compensation method is cascode compensation [6.3].

A differential-mode half-circuit for the operational amplifier is shown in Figure 6.7(a). The sign inversion represents the cross-coupling in the fully-differential circuit. If the current mirror is designed so that its response does not significantly influence the response of the amplifier at the frequencies of interest, and if the transconductances of M_5 and M_7 are equal, then the half-circuit of Figure 6.7(a) simplifies to that shown in Figure 6.7(b). With this simplification it is apparent that the circuit is basically just a two-stage amplifier with the second stage operating in push-pull fashion. The quiescent currents in the second stage are controlled by the common-mode output voltage of the first stage.

The common-mode half-circuit of the operational amplifier is shown in Figure 6.8. It is evident that the common-mode output of the first stage couples to the output of the second stage only through C_C. Therefore, two independent feedback circuits are needed to establish the dc common-mode levels at the outputs of the first and second stages. The common-mode output of the first stage is set to a level that establishes the desired quiescent currents in the second stage, while common-mode output of the sec-

Implementation of an Experimental Low-Voltage Modulator

Implementation of an Experimental Low-Voltage Modulator

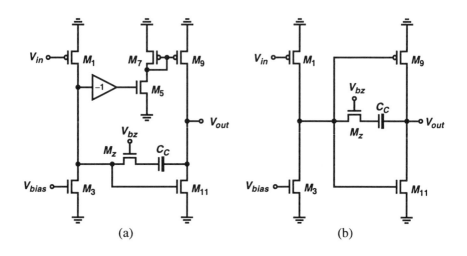

FIGURE 6.7 (a) Differential-mode half circuit, (b) simplified differential-mode half circuit.

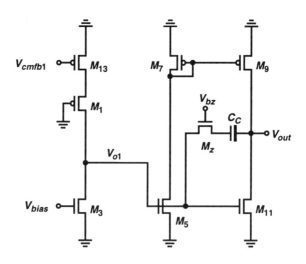

FIGURE 6.8 Common-mode half circuit

The Operational Amplifiers

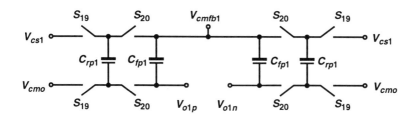

FIGURE 6.9 Common-mode feedback circuit for the first stage.

ond stage is set to midsupply to provide the maximum output swing. Two-stage class A amplifiers employing a single common-mode feedback loop require a sign inversion in the feedback loop that is commonly realized with a current mirror. To avoid compromising the stability of the common-mode feedback path, this current mirror must usually dissipate a significant amount of power. In the two-stage class A/AB amplifier the two common-mode feedback loops each include only a single stage of gain, and the power dissipation of the current mirror is avoided. The two common-mode feedback loops thus actually require less current than the single loop in a two-stage class A design.

The common-mode feedback circuit for the first amplifier stage is shown in Figure 6.9. V_{cs1} is the bias voltage for the tail current source in the amplifier. V_{cmo} is the desired common-mode output voltage of the first stage, which is set by a bias circuit that uses a scaled replica of M_{12} to establish the desired quiescent currents in the second stage. Since the drain current and W/L ratio are larger for M_{13} than for the input transistors, M_1 and M_2, the gain and bandwidth of the first stage common-mode feedback loop are greater than those of the first stage of the differential mode signal path.

The second stage common-mode feedback circuit, shown in Figure 6.10, senses the common-mode output voltage with a switched-capacitor network similar to that used in the first stage common-mode feedback loop. The output common-mode voltage is then used to control two common-source amplifiers, M_{B1} and M_{B3}, that drive the output nodes.

Switches S_{19} and S_{20} in the circuits of Figures 6.9 and 6.10 are controlled by $\Phi 1D$ and $\Phi 2D$, respectively, and are sized as 1.6 μm/0.8 μm. These small switch sizes minimize common-mode disturbances brought about by clock feedthrough and switch charge injection.

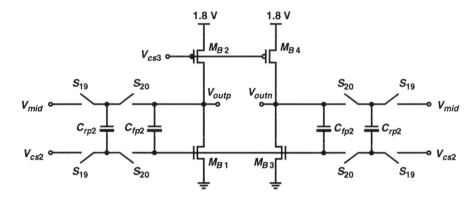

FIGURE 6.10 Common-mode feedback circuit for the second stage.

Each of the three operational amplifiers used to implement the integrators of Section 6.1 employs bias voltage generator circuits of the form shown in Figure 6.11. The bias voltages, V_{bias}, V_{cs1}, V_{cs2} and V_{cs3}, are used in the amplifier core, shown in Figure 6.6, as well as in the common-mode feedback circuits of Figures 6.9 and 6.10. The currents in the biasing circuits are set by off-chip resistors, R_{ref1} and R_{ref2}. The input common-mode voltage, V_{cmi}, and the mid supply reference, V_{mid}, are also generated with off-chip resistive voltage dividers. The off-chip circuitry used with the experimental modulator is discussed in Appendix D.

6.2.1 First-Stage Operational Amplifier

Transistor sizing in the operational amplifiers is governed by the need to accommodate a 1.8-V supply. As stated earlier, the amplifiers are designed for use with a low common-mode input voltage of 400 mV. The overdrive voltage (difference between the gate-to-source voltage and the threshold voltage) of the input transistors is 150 mV, while that of the tail current source is 100 mV. The quiescent voltage at the tail node is approximately 1.4 V. This ensures that transient voltage spikes at the amplifier's inputs do not force the tail current source out of the saturation region, which could result in a long settling response. The saturation voltage of the amplifier's output transistors is 100 mV. Thus, the largest amplifier output swing of 70% of the supply voltage (1.26 V), as predicted in Section 5.2, still leaves the output transistors well within the saturation region. It should be noted that variations in the ampli-

The Operational Amplifiers

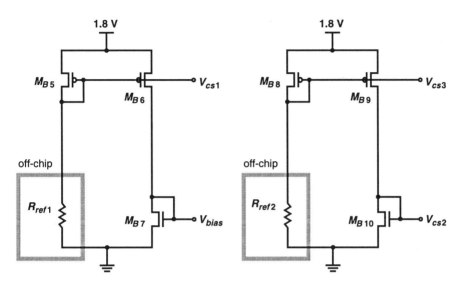

FIGURE 6.11 Biasing circuits.

fier's open loop gain can be significant when the drain-to-source voltage of the output transistors becomes less than approximately twice the saturation voltage.

To make the thermal noise of the first-stage operational amplifier approximately equal to the kT/C noise of the first integrator when both are referred to the input of the modulator, the compensation capacitor, C_C, is set to 4 pF. To meet the slew rate requirement in the first amplifier, as given by (5.49), with $C_C = 4$ pF, the tail current must be at least 50 µA. A current of 64 µA was chosen to provide a margin of safety. To meet the requirement for the settling time constant given by (5.50), the input transistor W/L ratio was chosen to be 120. To achieve an open loop dc gain of greater than 60 dB, a drawn channel length of 1.2 µm was used for the input transistors, which results in an electrical channel length of 1 µm. It follows from (5.42) that in order to limit the contribution of the load transistors in the first stage to the input-referred thermal noise of the amplifier, their g_m must be small compared to that of the input transistors. Therefore, these devices are sized much smaller than the input devices.

To ensure adequate phase margin, the quiescent current in transistors M_9 to M_{12} in the operational amplifier is 75 µA. Although the current-mirror pole would have not

Implementation of an Experimental Low-Voltage Modulator

TABLE 6.7 Transistor sizes in the first amplifier.

Transistor	Size (μm)	Transistor	Size (μm)
M_1, M_2	120/1.2	M_3, M_4	8/2
M_5, M_6, M_{11}, M_{12}	160/0.8	M_7, M_8, M_9, M_{10}	332.8/0.8
M_{13}	393.6/0.8	M_z	28.8/0.8
M_{B1}, M_{B3}, M_{B10}	21.6/0.8	M_{B2}, M_{B4}	48/0.8
$M_{B5}, M_{B6}, M_{B8}, M_{B9}$	48/0.8	M_{B7}	2/2

TABLE 6.8 Capacitor sizes in the first amplifier.

Capacitor	Size
C_C	4 pF
C_{fp1}, C_{fn1}	400 fF
C_{rp1}, C_{rn1}	200 fF
C_{fp2}, C_{fn2}	200 fF
C_{rp2}, C_{rn2}	100 fF

interfered with the operation of the second stage even with a current gain of two, a conservative current gain of one was used.

Simulations predict an open loop dc gain of 69 dB for this amplifier. The transistor sizes for the first amplifier are summarized in Table 6.7, and the capacitor sizes are given in Table 6.8.

The feedback capacitors in the common-mode feedback circuits, C_{fp1}, C_{fn1}, C_{fp2} and C_{fn2}, and the parasitic capacitance at the gates of the common-mode feedback transistors, M_{13}, M_{B1} and M_{B3}, form a voltage divider that attenuates the common-mode feedback signal. Unfortunately, the feedback capacitors also load the differential-mode signal path of the amplifier. To avoid excessive attenuation in the common-mode signal path without significantly increasing the loading on the differential-mode signal path, the feedback capacitors are sized so that they are comparable to the transistor parasitics. The size of the refresh capacitors, C_{rp1}, C_{rn1}, C_{rp2} and C_{rn2}, determines the number of clock cycles needed to establish the desired voltages on the level-shifting feedback capacitors. Also, as charge is transferred from the refresh capacitors to the feedback capacitors, there is a slight disturbance in the output common-mode levels.

6.2.2 Second-Stage Operational Amplifier

The dc voltages in the second-stage operational amplifier are approximately the same as those in the first-stage amplifier. However, since the capacitances in the second integrator are smaller and the settling requirements are not as demanding as in the first integrator, the transistor sizes and currents are scaled down in the second-stage amplifier.

The size of the compensation capacitor, C_C, in the second operational amplifier is 1.5 pF so that its associated thermal and kT/C noise contributions are approximately equal. To meet the slew rate requirement stated in (5.51) with this C_C, a tail current of 24 µA is needed. To meet the requirement for the settling time constant in (5.52), the input transistor W/L ratio was chosen to be 60. To achieve an open loop dc gain greater than 60 dB, a channel length of 1.2 µm was used for the input transistors. The quiescent current in the second stage of this amplifier is 48 µA. The transistor sizes for the second amplifier are summarized in Table 6.9, and the capacitor sizes are given in Table 6.10.

TABLE 6.9 Transistor sizes in the second amplifier.

Transistor	Size (µm)	Transistor	Size (µm)
M_1, M_2	60/1.2	M_3, M_4	2.4/2
M_5, M_6, M_{11}, M_{12}	25.6/0.8	M_7, M_8, M_9, M_{10}	80/0.8
M_{13}	108.8/0.8	M_z	8/0.8
M_{B1}, M_{B3}, M_{B10}	6.4/0.8	M_{B2}, M_{B4}	28.8/0.8
M_{B5}, M_{B9}	28.8/0.8	M_{B6}	57.6/0.8
M_{B7}	2.4/2	M_{B8}	12.4/0.8

Implementation of an Experimental Low-Voltage Modulator

TABLE 6.10 Capacitor sizes in the second amplifier.

Capacitor	Size
C_C	1.5 pF
C_{fp1}, C_{fn1}	150 fF
C_{rp1}, C_{rn1}	100 fF
C_{fp2}, C_{fn2}	100 fF
C_{rp2}, C_{rn2}	100 fF

6.2.3 Third-Stage Operational Amplifier

The dc voltages in the third-stage operational amplifier are approximately the same as those in the amplifiers used in the first two stages. The feedback factor in the third integrator is smaller than that in the second integrator, thereby increasing the circuit's settling time. However, since the settling requirements are less demanding for this integrator, the currents can still be scaled below those used in the second integrator stage. Since the baseband thermal noise of the third amplifier is greatly attenuated by noise shaping when referred to the input of the modulator, the size of the compensation capacitor is not critical, and 1 pF was used. A tail current of 16 µA meets the slew rate requirement given in (5.53) when $C_C = 1$ pF. To meet the requirement for the settling time constant in (5.54), the input transistor W/L ratio was chosen to be 33. A quiescent current of 24 µA in the second stage provides adequate phase margin. The transistor sizes for the third amplifier are given in Table 6.11, and the capacitor sizes are given in Table 6.12.

TABLE 6.11 Transistor sizes in the third amplifier.

Transistor	Size (µm)	Transistor	Size (µm)
M_1, M_2	20/0.8	M_3, M_4	1.6/2
M_5, M_6, M_{11}, M_{12}	12.8/0.8	M_7, M_8, M_9, M_{10}	40/0.8
M_{13}	54.4/0.8	M_z	4/0.8
M_{B1}, M_{B3}, M_{B10}	6.4/0.8	$M_{B2}, M_{B4}, M_{B6}, M_{B9}$	28.8/0.8
M_{B5}, M_{B8}	14.4/0.8	M_{B7}	1.6/2

TABLE 6.12 Capacitor sizes in the third amplifier.

Capacitor	Size
C_C	1 pF
C_{fp1}, C_{fn1}	150 fF
C_{rp1}, C_{rn1}	100 fF
C_{fp2}, C_{fn2}	100 fF
C_{rp2}, C_{rn2}	100 fF

6.3 The Quantizers

The two quantizers in the experimental 2-1 cascaded modulator each comprise a comparator and a 1-bit D/A converter. The comparator is implemented as a cascade of a regenerative latch followed by an SR latch, while the 1-bit DAC is a simple switch network connected to off-chip reference voltages. Errors introduced by the DAC in the first stage of the modulator are added to the input signal and directly degrade the performance of the modulator. Therefore, this DAC is designed to settle to the resolution of the modulator. However, errors associated with the other components of the two quantizers are greatly attenuated in the baseband by noise shaping. Thus, the principle design objective for these circuits is low power consumption.

6.3.1 Comparators

The two 1-bit quantizers are realized using the low power, dynamic latch illustrated in Figure 6.12 [6.13], followed by a simple static SR latch. The transistor sizes for the circuit in Figure 6.12 are given in Table 6.13. The dynamic latch operates as follows. When the clock $\Phi 1E$ is low, the latch is reset, and its outputs are both pulled high. Then, when $\Phi 1E$ goes high, the latch enters the regeneration phase. The cross-coupled transistors, M_3, M_4, M_7 and M_8, form a positive feedback loop and amplify the difference in the inputs to a full-rail output. Transistors M_3 and M_4 also serve to break the current path between the supplies once a decision has been made, and they provide some measure of isolation of the inputs from kickback during regeneration. The outputs of the dynamic latch are stored in an SR latch that is implemented with CMOS NAND gates.

Implementation of an Experimental Low-Voltage Modulator

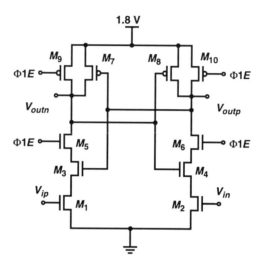

FIGURE 6.12 Regenerative comparator.

TABLE 6.13 Transistor sizes in the regenerative latches.

Transistor	Size (μm)	Transistor	Size (μm)
M_1, M_2	12.8/0.8	M_3, M_4	8/0.8
M_5, M_6	8/0.8	M_7, M_8	24/0.8
M_9, M_{10}	8/0.8		

In the architecture of Figure 6.1, the output of the second integrator is sampled by both the first-stage comparator and the third integrator. The sampling operation at the input of the third integrator causes a brief disturbance at the output of the second integrator. If the regenerative latch in the first stage quantizer enters the regeneration phase before the output of the second integrator has recovered, an erroneous result may be produced. To avoid this problem, an early phase of $\Phi 1$, labeled $\Phi 1E$, is used to strobe the first-stage comparator. The timing of the phases and the speed of the latch are such that the subsequent sampling by the third integrator does not cause the latch to err.

To minimize coupling from the comparator outputs into the analog circuits, the outputs of the two comparators are driven off chip with open-drain differential pairs. The resulting output currents are converted to voltages using off-chip components and are subsequently recorded by a data acquisition system. The test setup is described in greater detail in Appendix D.

6.3.2 Digital-to-Analog Converters

The 1-bit D/A converters are implemented as switch networks that are controlled by the comparator outputs. The outputs of the SR latches used in the comparators are buffered by inverters that drive NMOS switches connected to the negative reference voltage and PMOS switches connected to the positive reference voltage. The switches are sized so that the settling time constant for both the first and second stage DAC's is 2 ns.

The experimental modulator design employs feedback reference voltages equal to the supply voltages since lowering the feedback levels reduces the dynamic range of the converter. The reference voltages must be conditioned so that they do not introduce noise into the modulator. If the references are generated on chip, a supply-independent circuit that can provide a voltage close to V_{DD} should be used.

6.4 The Clocks

To digitize digital-audio band signals at a Nyquist sampling rate of 50 kHz, the experimental modulator is clocked at the rate of 4 MHz using seven clock phases, as noted in Section 6.1. Since the settling time of the integrator circuits is shorter during the sampling phase than the charge transfer phase, clocking with 50% duty cycle does not take full advantage of the circuit speed during the sampling phase. In order to experiment with the duty cycle, the phase 1 and phase 2 clocks were brought on chip separately. Experimentation showed that a duty cycle of 33% for phase 1 and 60% for phase 2 yields the best performance. The remaining 7% of the clock period is used to separate the different clock edges so as to avoid overlap among the phases.

Several clocking alternatives should be considered in future work. First, note that only the falling edges of the clocks need be nonoverlapping in order to avoid signal dependent switch charge injection. Therefore, a more sophisticated clocking circuit that creates synchronous rising edges but separated falling edges could be employed to

Implementation of an Experimental Low-Voltage Modulator

Implementation of an Experimental Low-Voltage Modulator

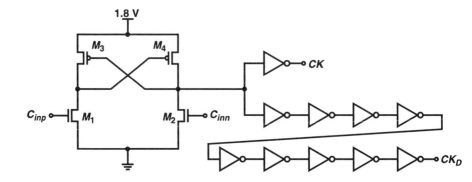

FIGURE 6.13 Clock generator.

increase the available settling time. Furthermore, dedicating 7% of the clock period to avoiding overlap among the clock phases is excessive and can be reduced. Finally, an alternative clocking strategy known as double sampling might be employed [6.4]. In this technique, the integrators sample and integrate on every half cycle, thus potentially doubling the throughput. However, this approach increases the loading on the integrators, thereby increasing their settling time beyond that obtainable with conventional clocking strategies. Furthermore, when double sampling is employed, mismatch between two nominally identical sampling capacitors can result in significant performance degradation.

6.4.1 Clock Generators

The clock generation circuit used in the experimental prototype is shown in Figure 6.13. One such circuit was used to create the phase 1 clocks and another to create the phase 2 clocks. To minimize the coupling of clocks through the package pins into the input signal and bias voltages, the external clocks are brought on chip as differential signals with an amplitude of 400 mV. The amplitude of the clocks is restored to full-rail on chip using common-source amplifiers with cross coupled loads. The chain of inverters generates the delay needed to separate the delayed clock phases from the nominal phases.

The outputs of the clock generators are fed to clock bootstrapping circuits to create the boosted clock levels and to generate the complementary phases. The nondelayed output of the phase 1 clock generator is also used as $\Phi 1E$ to strobe the regenerative latch. The bootstrapping circuits are discussed in the following section.

Clock Boosters

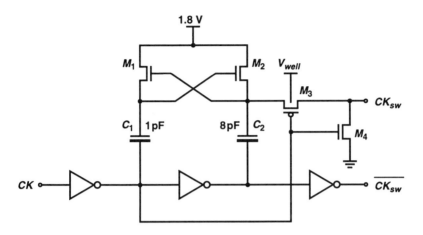

FIGURE 6.14 Clock boosting circuit.

6.5 Clock Boosters

In 5-V circuits, rail-to-rail operation of analog switches can be achieved through the use of CMOS transmission gates in which the NMOS and PMOS devices are driven by complementary control signals. The maximum on-resistance of a CMOS switch occurs when the input level is near midsupply. As the supply voltage is reduced, this maximum on-resistance increases quickly. In order to guarantee an adequately low switch resistance in a low voltage environment, the clock voltage used to drive at least one of the two switch transistors can be bootstrapped beyond the supply voltage range.

Shown in Figure 6.14 is a circuit that can be used to boost the signal driving the NMOS switch transistors above V_{DD} [6.13]. Capacitors C_1 and C_2 are charged to V_{DD} via the cross-coupled NMOS transistors M_1 and M_2. When the input clock, CK, goes high, the output voltage, CK_{sw}, approaches $2V_{DD}$. The output voltage does not actually reach $2V_{DD}$ because of charge sharing with the switch gates and parasitic capacitances to ac ground. To avoid charge sharing with the large well parasitic, and to reduce the potential for latch-up, the well of the PMOS transistor M_3 is tied to an on-chip voltage doubler. Capacitor C_1 can be relatively small as it only drives the gate of a single NMOS transistor, M_2. However, capacitor C_2 must be large enough to boost the gates of many NMOS switch transistors, as well as wiring parasitics. In the experimental modulator, a boosted clock voltage of 3.3 V was obtained when operating

Implementation of an Experimental Low-Voltage Modulator   **137**

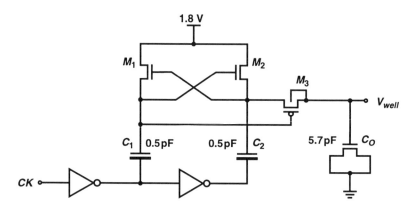

FIGURE 6.15 Voltage doubler.

from a 1.8-V power supply. Four of these circuits were used to generate the clock phases $\Phi1$, $\Phi1D$, $\overline{\Phi1D}$, $\Phi2$, $\Phi2D$, $\overline{\Phi2D}$. The inverter that generates the complement of CK_{sw} is used only in the clock boosters for $\Phi1D$ and $\Phi2D$ since these are the only clock phases that require the complementary phases.

6.5.1 Voltage Doublers

The well of the PMOS switch, M_3, in Figure 6.14 is biased by the circuit shown in Figure 6.15 [6.13]. This circuit produces a voltage of $2V_{DD}$ while consuming only a few microwatts of power. The bypass capacitor, C_O, is formed with an NMOS transistor and serves to reduce the ripple in the output voltage. A similar circuit is used to generate the voltage, V_{bz}, that is used that is used in the amplifiers to bias the compensation transistors, M_z in Figure 6.6. To minimize the ripple in V_{bz}, $C_1 = C_2 = 0.2\,\text{pF}$ and $C_O = 40\,\text{pF}$ in the compensation bias circuit.

6.6 Decimation Filtering

Hardware implementations of digital decimation filters for oversampling modulators can be efficiently implemented with a cascade of sinc filters that reduce the sampling frequency to within a factor of four of the Nyquist rate, followed by FIR or IIR filters that provide the final factor of four decimation [6.5]. A sinc filter computes a running

Experimental Results

average of its inputs and must perform division by its decimation ratio. When the decimation ratio of the sinc filter is an integer power of 2, the division is easily implemented with a binary shifter.

Division by integers that are not a power of 2 can be approximated efficiently using

$$a \sum_{n=0}^{\infty} r^n = a + ar + ar^2 + \ldots, \qquad (6.3)$$

the sum of which, when $|r| < 1$, is

$$S = \frac{a}{1-r}. \qquad (6.4)$$

In this work an oversampling ratio of 80 is used, which requires decimation by 20 in the sinc filters. Division by 20 can be performed efficiently in hardware by using (6.3) with $a = 1/16$ and $r = -1/4$. This results in the following sum:

$$\sum_{n=2}^{\infty} \left(\frac{-1}{4}\right)^n = \frac{1}{16} - \frac{1}{64} + \frac{1}{256} - \ldots = \frac{1}{20}. \qquad (6.5)$$

6.7 Experimental Results

The performance of the modulator in Figure 6.1 was assessed by driving it with a fully-differential sinusoidal signal, acquiring the digital outputs from the two stages with a custom test board, then transferring the data to a workstation for subsequent signal processing. Digital filtering and signal analysis were performed using the program MIDAS [6.6], [6.7].

A die micrograph of the experimental modulator is shown in Figure 6.16. The clocks are brought on chip as differential, low-swing voltages, and the two 1-bit digital outputs are driven off chip as differential currents. Separate digital and analog ground planes were used on the test board and tied together at the power supply. The feedback reference voltages were brought on chip via dedicated pins and tied to analog power and ground off chip. The common-mode level of the input source was tied to the midsupply voltage of the analog supplies, and the analog input, the feedback ref-

Implementation of an Experimental Low-Voltage Modulator

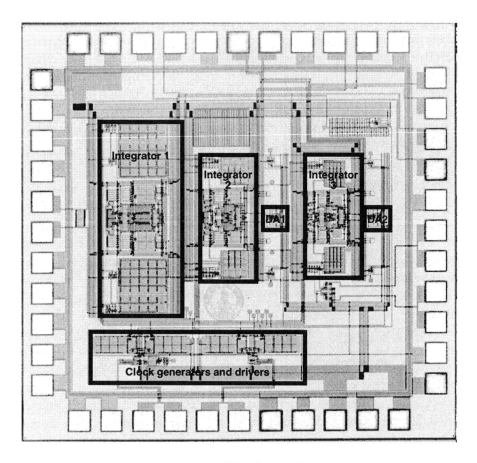

FIGURE 6.16 Die micrograph.

erence voltages, and the integrator outputs were all referenced to that midsupply voltage. The test setup is described in Appendix D.

When operated from a 1.8-V supply, the modulator's power dissipation, excluding the output drivers, is 2.5 mW. At a sampling frequency of 4 MHz with an oversampling ratio of 80, the corresponding Nyquist sampling rate is 50 kHz, and the signal bandwidth is 25 kHz. Plots of the *SNR* and *SNDR* versus input amplitude for an input sinusoid at approximately 2 kHz are shown in Figure 6.17. To generate these plots, the amplitude of the sinusoidal signal source was stepped in 1-dB increments from 0 dB to −85 dB. At each step, an FFT was taken, and the *SNR* and *SNDR* were recorded.

Experimental Results

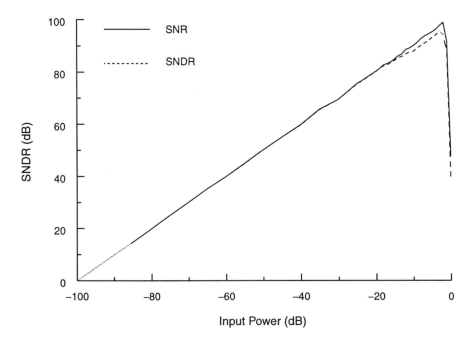

FIGURE 6.17 Measured SNR and SNDR vs. input power.

The curves are extrapolated from −85 dB to the noise floor of the modulator. The overload level of the modulator is −1 dB below the differential reference voltage of $2V_{DD}$. Thus, the differential input range is 3.2 V. From Figure 6.17 it is apparent that for digital-audio bandwidths the modulator achieves a dynamic range of 99 dB and a peak *SNDR* of 95 dB when operated from a 1.8-V supply. The measured performance of the modulator is summarized in Table 6.14.

Figure 6.18 plots an FFT of the output spectrum for a −20-dB, 2-kHz input signal. The frequency independence of the noise floor indicates that the modulator's performance is limited by thermal, rather than quantization, noise. Figure 6.19 plots dynamic range as a function of oversampling ratio for a sampling rate of 4 MHz. While in a quantization noise-limited third-order modulator the dynamic range increases by 21 dB/octave increase in oversampling ratio, in the thermal noise limit the increase is only 3 dB/octave. It is apparent from Figure 6.19 that for oversampling ratios below 64, the experimental modulator is quantization noise limited, while for oversampling ratios above 64 its performance is limited by thermal noise.

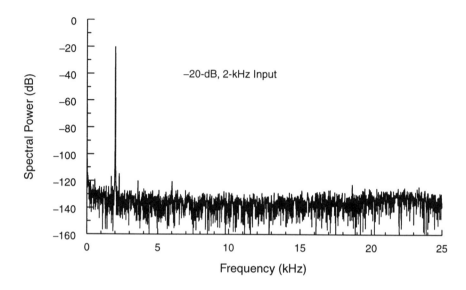

FIGURE 6.18 Measured baseband output spectrum.

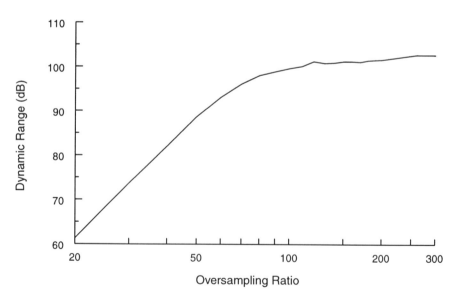

FIGURE 6.19 Measured dynamic range vs. oversampling ratio.

Experimental Results

TABLE 6.14 Performance summary.

Dynamic Range	99 dB
Peak SNR	99 dB
Peak SNDR	95 dB
Noise Floor	−100 dB
Overload Level	−1 dB
Sampling Rate	4 MHz
Oversampling Ratio	80
Signal Bandwidth	25 kHz
Power Supply Voltage	1.8 V
Power Dissipation	2.5 mW
Active Area	1.5 mm^2
Technology	0.8-μm CMOS, 1-poly 2-metal with poly-to-metal capacitors

Figure 6.20 plots the measured dynamic range and power dissipation as a function of the supply voltage for a signal bandwidth of 25 kHz and an oversampling ratio of 80. The modulator operates successfully for supply voltages ranging from 1.5 V to 2.5 V. The minimum supply voltage is limited by the input stage of the operational amplifiers, while the maximum supply voltage operation is limited by the oxide stress produced by the voltage doublers. The dynamic range improves with higher supply voltages due to the increased maximum signal power and a constant thermal noise floor. Below a supply voltage of 1.6 V, the modulator's resolution falls off quickly due to charge loss at the integrator inputs brought about by a slight forward biasing of the switch source/drain junctions that results from transient voltage spikes. Since the bias currents are held constant and the power dissipation is dominated by the analog circuits, the power dissipation increases linearly with supply voltage.

Figure 6.21 plots the common-mode rejection ratio (*CMRR*) of the experimental modulator across the Nyquist bandwidth. A common-mode input sinusoid with an amplitude of −3 dB was applied at several frequencies, and the power of the resulting tone in the output of the modulator was recorded. The *CMRR* can then be computed by subtracting the power of the tone that appears in the output of the modulator in response to a common-mode input from −3 dB. This measurement shows that a high *CMRR* can be achieved even in a low-voltage environment when the class A/AB operational amplifier is used.

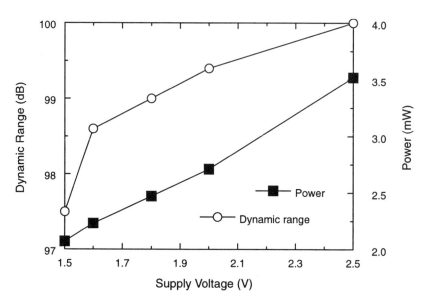

FIGURE 6.20 Measured dynamic range and power dissipation vs. supply voltage.

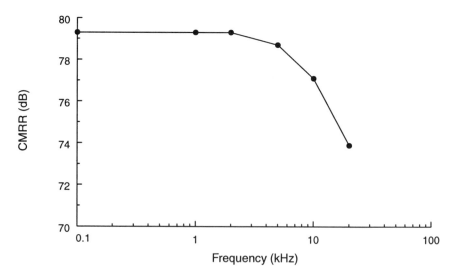

FIGURE 6.21 Measured common-mode rejection ratio.

Experimental Results

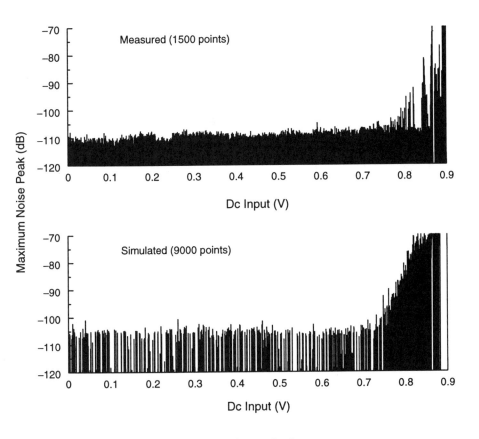

FIGURE 6.22 Measured quantization tones.

An important phenomenon that can degrade the performance of $\Sigma\Delta$ modulators is the presence of spurious noise tones in the baseband output spectrum [6.5]. Figure 6.22 plots the maximum baseband tone peak that results from a dc input as a function of the dc input level. To generate the plots of Figure 6.22, the dc input was stepped in fine increments over the input range from zero to the maximum positive input for both the experimental modulator and a simulation. Results for the negative half of the input range are symmetric about the zero input. At each dc input value, an FFT was taken and the power of the maximum baseband noise peak was recorded. For dc levels near the zero input, the output spectrum is free of tones. For dc inputs below −1.1 dB of full scale, the strongest tone in the measured data is below the thermal noise floor of 100 dB.

Implementation of an Experimental Low-Voltage Modulator **145**

The power of tones plotted in Figure 6.22 is lower in the experimental measurements than in the simulation. In simulations, relatively strong tones occur at certain dc levels, while tones at neighboring dc levels may be quite low. However, the small amount of thermal noise present in the experimental modulator spreads the tones across a dc input range, thereby avoiding the tone energy that is concentrated at a particular dc level.

6.8 Comparison of the Power Efficiency of A/D Converters

The power efficiency of various A/D converters with different resolutions and Nyquist sampling rates can be compared using the following figure of merit,

$$FM = \frac{4kT \times DR \times f_N}{P},\qquad(6.6)$$

where P is the total power dissipation of the converter, f_N is its Nyquist sampling rate, and DR is expressed as a power ratio rather than in dB. This figure of merit is defined so that $FM = 1$ for a single differential, switched-capacitor integrator implemented with an ideal class B operational amplifier. Figure 6.23 plots FM for some recently published CMOS A/D converters [6.8], [6.9], [6.10]-[6.14], [6.15], [6.16], [6.17]-[6.23]. The power dissipation used to compute FM does not include the decimation filter in the case of oversampling converters, nor does it account for the antialias filter in Nyquist-rate converters. As expected from the discussion in Section 4.3, the dependence of the figure of merit on conversion rate becomes quadratic for faster, lower resolution converters.

6.9 Summary

The circuits presented in this chapter were used to implement the modulator architecture established Chapter 5 in a manner that would allow operation from a power supply as low as 1.5 V. Throughout the design, robustness was emphasized over power conservation. Therefore, the power dissipation in the second and third integrators was not scaled aggressively. Nevertheless, the power dissipation of 1.6 mW in the analog

Summary

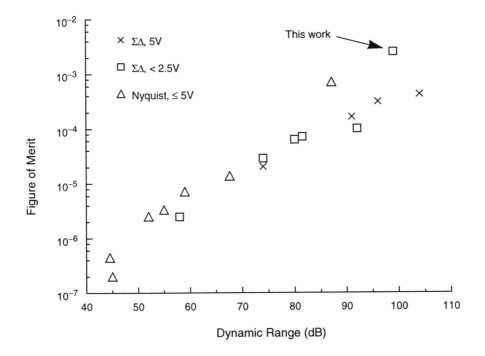

FIGURE 6.23 Figure of merit vs. dynamic range of recent CMOS A/D converters.

circuits and 0.9 mW in the associated digital circuits represents a significant improvement over previously reported results.

REFERENCES

[6.1]   K. L. Lee and R. Meyer, "Low-distortion switched-capacitor filter design techniques," *IEEE J. Solid-State Circuits,* vol. SC-20, pp. 1103-1112, December 1985.

[6.2]   W. Black, D. Allstot, and R. Reed, "A high performance low power CMOS channel filter," *IEEE J. Solid-State Circuits,* vol. SC-15, pp. 929-938, Dec. 1980.

[6.3] B. Ahuja, "An improved frequency compensation technique for CMOS operational amplifiers," *IEEE J. Solid-State Circuits*, vol. SC-18, pp. 629-633, Dec. 1983.

[6.4] D. Senderowicz, et al., "Low voltage double sampled ΣΔ converters," *IEEE J. Solid-State Circuits*, vol. SC-32, pp. 1907-1919, Dec. 1997.

[6.5] J. Candy and G. Temes, "Oversampling methods for A/D and D/A conversion," in *Oversampling Delta-Sigma Data Converters*, pp. 1-29, New York: IEEE Press, 1992.

[6.6] L. Williams, and B. A. Wooley, "MIDAS – a functional simulator for mixed digital and analog sampled data systems," *Proc. 1992 IEEE Int. Symp. Circuits Syst.*, pp. 2148-2151, May 1992.

[6.7] S. Rabii, L. Williams, B. Boser, and B. A. Wooley, *MIDAS User Guide Version 3.1*, Stanford University, Stanford, CA, 1997.

[6.8] E. J. van der Zwan and E. C. Dijkmans, "A 0.2mW CMOS ΣΔ modulator for speech coding with 80dB dynamic range," *ISSCC Digest of Tech. Papers*, pp. 232-233, February 1996.

[6.9] S. Kiriaki, "A 0.25mW sigma-delta modulator for voice-band applications," *Symp. on VLSI Circuits Digest of Tech. Papers*, pp. 35-36, 1995.

[6.10] J. Grilo, E. MacRobbie, R. Halim, and G. Temes, "A 1.8V 94dB dynamic range ΣΔ modulator for voice applications," *ISSCC Digest of Tech. Papers*, pp. 232-233, February 1996.

[6.11] K. Kusumoto, A. Matsuzawa, and K. Murata, "A 10-b 20-MHz 30-mW pipelined interpolating CMOS ADC," *IEEE J. Solid-State Circuits*, vol. SC-28, pp. 1200-1206, December 1993.

[6.12] M. Yotsuyanagi, H. Hasegawa, M. Yamaguchi, M. Ishida, and K. Sone, "A 2V 10b 20MSample/s mixed-mode subranging CMOS A/D converter," *ISSCC Digest of Tech. Papers*, pp. 282-283, February 1995.

[6.13] T. B. Cho and P. R. Gray, "A 10 b, 20 Msample/s, 35 mW pipeline A/D converter," *IEEE J. Solid-State Circuits*, vol. SC-30, pp. 166-172, March 1995.

Summary

[6.14]   A. G. W. Venes and R. J. van de Plassche, "An 80MHz 80mW 8b CMOS folding A/D converter with distributed T/H preprocessing," *ISSCC Digest of Tech. Papers*, pp. 318-319, February 1996.

[6.15]   L. Williams and B. A. Wooley, "A third-order sigma-delta modulator with extended dynamic range," *IEEE J. Solid-State Circuits*, vol. SC-29, pp. 193-202, March 1994.

[6.16]   B. P. Brandt and B. A. Wooley, "A 50-MHz multibit sigma-delta modulator for 12-b 2-MHz A/D conversion," *IEEE J. Solid-State Circuits*, vol. SC-26, pp. 1746-1756, December 1991.

[6.17]   G. Yin, F. Stubbe, and W. Sansen, "A 16-b 320-kHz CMOS A/D converter using two-stage third-order $\Sigma\Delta$ noise shaping," *IEEE J. Solid-State Circuits*, vol. SC-28, pp. 640-647, June 1993.

[6.18]   P. C. Yu and H.-S. Lee, "A 2.5V 12b 5MSample/s pipelined CMOS ADC," *ISSCC Digest of Tech. Papers*, pp. 314-315, February 1996.

[6.19]   M. P. Flynn and D. J. Allstot, "CMOS folding ADCs with current-mode interpolation," *ISSCC Digest of Tech. Papers*, pp. 374-275, February 1995.

[6.20]   Y. Matsuya and J. Yamada, "1 V power supply, low power consumption A/D conversion technique with swing-suppression noise shaping," *IEEE J. Solid-State Circuits*, vol. SC-29, pp. 1524-1530, December 1994.

[6.21]   V. Peluso, M. Steyaert, and W. Sansen, "A 1.5-V 100µW $\Delta\Sigma$ modulator with 12-b dynamic range using the switched-opamp technique," *IEEE J. Solid-State Circuits*, vol. SC-32, pp. 943-952, July 1997.

[6.22]   S.-U. Kwak, B.-S. Song, and K. Bacrania, "A 15-b 5-Msample/s low-spurious CMOS ADC," *IEEE J. Solid-State Circuits*, vol. SC-32, pp. 1866-1875, December 1997.

[6.23]   A. Marques, V. Peluso, M. Steyaert, and W. Sansen, "A 15-b resolution 2-MHz Nyquist rate $\Delta\Sigma$ ADC in a 1-µm CMOS technology," *IEEE J. Solid-State Circuits*, vol. SC-33, pp. 1065-1075, July 1998.

CHAPTER 7 *Conclusion*

Sigma-delta ($\Sigma\Delta$) modulation has been shown to be an especially effective means of implementing high-resolution analog-to-digital converters. This work has sought to demonstrate the feasibility of using such an approach in applications where the supply voltage is less than 2 V. The work has focused on four principle topics: the analysis of power dissipation in a $\Sigma\Delta$ modulator, a comparison of analog circuit topologies for low-voltage design, a comparison of modulator architectures for low-voltage applications, and the design of a high-resolution, low-voltage, low-power modulator.

An analysis of power dissipation identified the first integrator stage as consuming the most power among the circuit blocks that comprise a highly oversampled $\Sigma\Delta$ modulator. It was noted that lowering the oversampling ratio is of limited benefit in reducing the power dissipated in an oversampling modulator. It was also confirmed that a reduction in supply voltage generally results in an increase in power dissipation in high-resolution analog circuits.

A comparison of the power dissipated in a variety of circuit topologies, including switched-capacitor, switched-current and continuous-time, revealed that switched-capacitor circuits have the highest potential for low-power operation in high-resolution applications. A class A/AB operational amplifier was developed for use in low-voltage switched-capacitor circuits. This amplifier provides a large output signal range, high rejection of common-mode signals, and high power supply rejection, even when operating from a low supply voltage.

Conclusion

A study of ΣΔ modulator architectures identified the 2-1 cascaded topology as especially suitable for low-voltage and low-power design. An implementation of this architecture achieved both a high overload level relative to the supply voltage and excellent spectral purity. By employing signal scaling within the modulator, the output swings of the integrator blocks were designed to remain well within the linear output range of a practical integrator implementation.

An experimental modulator based on the analyses presented in this work was designed and integrated in a 0.8-μm CMOS technology. This circuit takes advantage of high swing class A/AB operational amplifiers, low-power regenerative comparators, and bootstrapped clock drivers to achieve a dynamic range of 99 dB over a 25-kHz bandwidth while dissipating only 2.5 mW from a 1.8-V supply.

7.1 Recommendations for Further Investigation

The work described herein may be extended to even lower voltage operation. As pointed out in Chapter 2, operation from a single battery cell will likely require a supply voltage of 0.9-1.0 V. Furthermore, to avoid the oxide stress that is likely to occur if bootstrapping is employed in aggressively scaled CMOS technologies, alternatives to the clock drivers shown in Section 6.5 should be explored. One candidate is the clock booster described in [7.1].

Additionally, the specifications for voiceband data acquisition, 80 dB dynamic range over a 4-kHz bandwidth, is of wide interest for various portable phones, while a signal bandwidth of 10 kHz and very low power dissipation are needed in hearing aids.

Another area of significant interest is high-resolution and high-speed data conversion for wireless applications. A modulator that provides 85-90 dB dynamic range for a signal bandwidth of 1-5 MHz when operated from 2.5-V or lower supply voltage would be useful in meeting the requirements of a number of emerging communications standards [7.2]. Due to the high sampling speeds required in such a circuit, only low oversampling ratios can be used. Architectural innovations will be required to achieve adequate suppression of quantization noise with a low oversampling ratio. Moreover, while the class A/AB operational amplifier developed in this work has proven adequate for a medium-speed and high-resolution design, it is not well-suited for high-speed applications due to its relatively low-frequency non-dominant poles. Thus, the design of a low-voltage operational amplifier for use in high-speed switched-capacitor circuits will be essential for such applications.

REFERENCES

[7.1] T. Brooks, et al., "A 16b ΣΔ pipeline ADC with 2.5MHz output data-rate," *ISSCC Digest of Tech. Papers*, pp. 208-209, February 1997.

[7.2] D. Cox, et al., "Universal digital portable radio communications," *Proceedings of the IEEE*, vol. 75, pp. 436-477, April 1987.

APPENDIX A *Fundamental Limits*

This appendix presents a derivation of the results given in Section 4.2 for the power dissipation in ideal integrators of the forms shown in Figures 4.2, 4.3 and 4.4. For the first two integrator topologies, implementation with an ideal class A amplifier is considered initially, followed by an analysis based on the use of an ideal class B amplifier. The amplifiers are assumed to be noiseless, have infinite gain and bandwidth, and introduce no parasitic capacitances. For the third integrator topology, the switched-current circuit in Figure 4.4 is analyzed. The MOS transistors are assumed to be ideal square-law devices with infinite output resistance, and the switches are treated as ideal switches in series with resistors.

A.1 Power in a Switched-Capacitor Integrator

This section examines the power dissipation in a switched-capacitor integrator of the form shown in Figure 4.2 when it is used as the first stage in a $\Sigma\Delta$ modulator. The dynamic range (DR) of a $\Sigma\Delta$ modulator for which the resolution is limited by the kT/C noise of the first switched-capacitor integrator stage, given in (4.3) and repeated here for convenience, is

$$DR = \frac{S_S}{S_{kT/C}} = \frac{V_{sw}^2 M C_S}{8kT}, \qquad \text{(A.1)}$$

Fundamental Limits

where V_{sw} is the amplitude of a full-scale sinusoidal input to the modulator. From this expression is follows that

$$C_S = \frac{8kT(DR)}{MV_{sw}^2}. \tag{A.2}$$

The average power dissipation in the integrator is

$$P = I_{amp}V_{DD}, \tag{A.3}$$

where I_{amp} is the average amplifier current and V_{DD} is the supply voltage. In a class A amplifier the quiescent amplifier current is also the maximum current that can be delivered to the load. Therefore, the quiescent current must be large enough to ensure that the load can be charged quickly enough to accommodate the largest expected output voltage step within the integration period. Thus,

$$I_{amp} = \frac{C_I \Delta V_{out}}{T_S/2}, \tag{A.4}$$

where ΔV_{out} is the largest differential step change in the output voltage, T_S is the sampling period, and it is assumed that the integration must be completed in half the clock period. ΔV_{out} is related to V_{sw} by

$$\Delta V_{out} = (V_{sw} + V_{ref})\frac{C_S}{C_I}, \tag{A.5}$$

where V_{ref} is the amplitude of the differential feedback reference voltage of the modulator. Also,

$$T_S = \frac{1}{f_S} = \frac{1}{Mf_N}, \tag{A.6}$$

where f_S is the sampling frequency, f_N is the Nyquist sampling rate, and M is the oversampling ratio.

By substituting (A.5) and (A.6) into (A.4), the expression for I_{amp} can be rewritten as

$$I_{amp} = 2C_S M f_N (V_{sw} + V_{ref}). \tag{A.7}$$

Power in a Switched-Capacitor Integrator

From (A.2), (A.3) and (A.7), it follows that the power dissipation in the integrator is given by

$$P = 16kT(DR)f_N \frac{V_{DD}(V_{sw} + V_{ref})}{V_{sw}^2}. \qquad \text{(A.8)}$$

If it is further assumed that $V_{sw} = V_{ref} = V_{DD}$, (A.8) reduces to (4.4), namely

$$P = 32kT(DR)f_N. \qquad \text{(A.9)}$$

In a class B amplifier, the quiescent power consumption is zero; power is dissipated only when the output voltage changes. The average power dissipation of a fully differential integrator implemented with an ideal class B amplifier is [A.1]

$$P = \frac{1}{2}C_I f_S V_{DD} \overline{\Delta V_{out}}, \qquad \text{(A.10)}$$

where $\overline{\Delta V_{out}}$ is the average differential output voltage step size, which is related to the average differential input voltage step size by

$$\overline{\Delta V_{out}} = \frac{C_S}{C_I} \times \overline{\Delta V_{in}}. \qquad \text{(A.11)}$$

The average differential input voltage step size is

$$\overline{\Delta V_{in}} = \overline{V_{in} - V_{ref}}, \qquad \text{(A.12)}$$

where V_{in} is the modulator's input and V_{ref} is the feedback reference voltage. The maximum full-scale V_{sw} is V_{ref}.

Substituting (A.2), (A.6), and (A.11) into (A.10) results in

$$P = 4kT(DR)f_N \frac{\overline{\Delta V_{in}} V_{DD}}{V_{sw}^2}. \qquad \text{(A.13)}$$

If it is further assumed that $V_{sw} = V_{DD}$, (A.13) reduces to (4.5), namely

Fundamental Limits

$$P = 4kT(DR)f_N \frac{\overline{\Delta V_{in}}}{V_{DD}}. \tag{A.14}$$

A.2 Power in a Continuous-Time Integrator

This section examines the power dissipation in a continuous-time integrator of the form shown in Figure 4.3. The dynamic range (DR) of a $\Sigma\Delta$ modulator for which the resolution is limited by the noise of the input resistors in a first stage employing a continuous-time integrator, given in (4.8), is

$$DR = \frac{S_S}{S_N} = \frac{V_{sw}^2}{16kTRf_N}. \tag{A.15}$$

From this expression it follows that

$$R = \frac{V_{sw}^2}{16kT(DR)f_N}. \tag{A.16}$$

The average power dissipation in the integrator is

$$P = I_{amp} V_{DD}, \tag{A.17}$$

where I_{amp} is the average amplifier current and V_{DD} is the supply voltage. In a class A amplifier, the quiescent amplifier current is also the maximum current that can be delivered to the load. In a continuous-time integrator, the quiescent current of the operational amplifier must be at least equal to the current delivered by the source. Thus,

$$I_{amp} = \frac{V_{sw}}{R} - 2G_f V_{ref}, \tag{A.18}$$

where $2G_f V_{ref}$ is the differential feedback reference current of the modulator.

Overload of the modulator occurs when $2G_f V_{ref} = V_{sw}/R$. Therefore, (A.18) can be rewritten as

Power in a Switched-Current Integrator

$$I_{amp} = 2\frac{V_{sw}}{R}. \quad \text{(A.19)}$$

Substituting (A.16) and (A.19) into (A.17) results in (4.9), namely

$$P = 32kT(DR)f_N \frac{V_{DD}}{V_{sw}}. \quad \text{(A.20)}$$

In a class B amplifier, the average amplifier current is given by

$$\overline{I_{amp}} = \overline{\Delta I_{in}} = \frac{\overline{\Delta V_{in}}}{R} = \frac{\overline{V_{in} - 2G_f R V_{ref}}}{R}, \quad \text{(A.21)}$$

Substituting (A.16) and (A.21) into (A.17) results in (4.10), namely

$$P = 16kT(DR)f_N \frac{\overline{\Delta V_{in}} V_{DD}}{V_{sw}^2}. \quad \text{(A.22)}$$

A.3 Power in a Switched-Current Integrator

This section examines the power dissipation in a switched-current integrator of the form shown in Figure 4.4. The baseband kT/C noise in the drain current of M_1 in Figure 4.4 is

$$S_{N,M1} = \frac{kT}{MC_{GS1}} \times g_{m1}^2, \quad \text{(A.23)}$$

where g_{m1} is the transconductance of M_1. g_{m1} is given by

$$g_{m1} = \frac{2I_D}{V_{GS} - V_T}, \quad \text{(A.24)}$$

where I_D is the drain current of M_1. When $I_D = I_1$, substituting (A.24) into (A.23) results in

Fundamental Limits

Fundamental Limits

$$S_{N,M1} = \frac{kT}{MC_{GS1}} \times \frac{4I_1^2}{(V_{GS}-V_T)^2}. \tag{A.25}$$

If the response of M_1 is modeled as having a time constant

$$\tau_{M1} = \frac{C_{GS1}}{g_{m1}}, \tag{A.26}$$

and if the duration of the sampling phase is

$$\frac{T_S}{2} = 2\tau_{M1}, \tag{A.27}$$

then substituting (A.24) and (A.26) into (A.27) and setting $I_D = I_1$ results in

$$\frac{T_S}{2} = \frac{C_{GS1}(V_{GS}-V_T)}{I_1}. \tag{A.28}$$

Note that (A.27) allows for only two settling time constants during the sampling interval. In practice, nonlinearities in the response of the circuit require more complete settling if harmonic distortion is to be avoided.

From (A.28) it follows that

$$C_{GS1} = \frac{T_S I_1}{2(V_{GS}-V_T)}. \tag{A.29}$$

Substituting (A.29) and (A.6) into (A.25) results in

$$S_{N,M1} = 8kTf_N \frac{I_1}{(V_{GS}-V_T)}. \tag{A.30}$$

The baseband kT/C noise in a modulator that employs the switched-current integrator of Figure 4.4 includes the drain current noise of M_2 in addition to that of M_1. The noise in the drain current of M_3 is spectrally shaped and is therefore ignored here. Thus, the total input-referred kT/C noise is approximately twice that given by (A.30)

Power in a Switched-Current Integrator

$$S_{kT/C} = 16kTf_N \frac{I_1}{(V_{GS} - V_T)}. \quad \text{(A.31)}$$

The power of the integrator input current signal is

$$S_S = \frac{I_{in}^2}{2}, \quad \text{(A.32)}$$

where I_{in} is the amplitude of a full-scale sinusoidal input to the modulator. When $I_{in} \ll I_1$, to first order it follows from the square-law model of an MOS transistor that

$$I_{in} \approx \frac{2I_1}{(V_{GS} - V_T)} V_{in}, \quad \text{(A.33)}$$

where V_{in} is the amplitude of the voltage swing at the gate of M_1. Substituting (A.33) into (A.32) and simplifying gives

$$S_S = \frac{2I_1^2 V_{in}}{(V_{GS} - V_T)}. \quad \text{(A.34)}$$

From (A.31) and (A.34), it follows that the dynamic range of the integrator is

$$DR = \frac{S_S}{S_{kT/C}} = \frac{I_1 V_{in}^2}{8kTf_N (V_{GS} - V_T)}. \quad \text{(A.35)}$$

(A.35) can be rearranged to provide an expression for I_1 as

$$I_1 = 8kT(DR)f_N \frac{V_{GS} - V_T}{V_{in}^2}. \quad \text{(A.36)}$$

The power dissipation in the switched-current integrator of Figure 4.4 is

$$P = 3I_1 V_{DD}, \quad \text{(A.37)}$$

Substituting (A.36) into (A.37) gives (4.16), namely

Fundamental Limits

Fundamental Limits

$$P = 24kT(DR)f_N \frac{(V_{GS} - V_T)V_{DD}}{V_{in}^2}. \quad \text{(A.38)}$$

REFERENCES

[A.1] R. Castello and P. Gray, "Performance limitations in switched-capacitor filters," *IEEE Trans. on Circuits and Systems II*, vol. CAS-32, pp. 865–876, September 1985.

APPENDIX B   *Power Dissipation vs. Supply Voltage and Oversampling Ratio*

In this appendix, the equations used to generate the curves in Figures 4.7 and 4.8 are derived. The operational amplifier circuits examined here are shown in Figure 4.6, and their key performance parameters are summarized in Table 4.1. The amplifiers are assumed to be embedded in the feedback configuration shown in Figure 4.5. First, a general expression is derived for the tail current of the input differential pair that is used as an input stage in all three of the amplifiers. Then, the total current and power dissipation are computed for each of the amplifiers.

The response of an operational amplifier that employs a differential pair as its input stage typically includes a slew limited region followed by a linear response region. The duration of the slew limited region is

$$T_{SL} = \frac{V_f - (V_{GS1} - V_{T1})}{S}, \quad \text{(B.1)}$$

where V_f is the asymptotic final value of the output voltage, $V_{GS1} - V_{T1}$ is the overdrive voltage of the input transistors, and S is the slew rate. The maximum value of V_f is

$$V_{f,max} = \frac{C_S}{C_I}(V_{sw} - V_{ref}). \quad \text{(B.2)}$$

If the linear settling regime is characterized by a single pole response, its duration is

Power Dissipation vs. Supply Voltage and Oversampling Ratio

$$T_{Lin} = a\tau, \quad (B.3)$$

where τ is the settling time constant, and a is a factor that ensures settling to within 0.5 LSB of the modulator. a can be expressed as

$$a = \ln\sqrt{2DR} = \frac{1}{2}(\ln 2 + \ln DR), \quad (B.4)$$

where DR is the dynamic range of the modulator.

The sampling time is the sum of the duration of the two regimes and is given by

$$\frac{T_S}{2} = T_{SL} + T_{Lin} = \frac{(C_S/C_I)(V_{sw} - V_{ref}) - (V_{GS1} - V_{T1})}{S} + \frac{1}{2}(\ln 2 + \ln DR)\tau. \quad (B.5)$$

Substituting (A.6) and the relationship between S and τ in (5.16) into (B.5) results in

$$\left[\frac{2(C_S/C_I)(V_{sw} - V_{ref}) - 2(V_{GS1} - V_{T1})}{(1 + (C_S + C_P)/C_I)(V_{GS1} - V_{T1})} + \ln 2 + \ln DR\right] M f_N \tau = 1. \quad (B.6)$$

The settling time constant for a single stage amplifier in the feedback configuration of Figures 4.5 is given in (4.20) as

$$\tau = \frac{C_L + C_S + C_P + (C_L/C_I)(C_S + C_P)}{g_m}. \quad (B.7)$$

This expression also holds for a two-stage operational amplifier with pole-splitting compensation when the compensation capacitor, C_C, is

$$C_C = \frac{C_I(C_S + C_P)}{C_I + C_S + C_P}. \quad (B.8)$$

Substituting (B.7) and (A.24) into (B.6) and rearranging results in

$$I_1 = M f_N (\ln 2 + \ln DR)(C_S + C_P)(V_{GS1} - V_{T1})$$
$$+ 2M f_N \frac{C_I(C_S + C_P)}{C_I + C_S + C_P}\left(\frac{C_S}{C_I}(V_{sw} - V_{ref}) - (V_{GS1} - V_{T1})\right), \quad (B.9)$$

where the tail current $I_1 = 2I_D$, and $C_L = 0$ because the parasitics associated with C_I and the output transistors are ignored in this idealized analysis.

The size of the sampling capacitor, C_S, is given in (A.2). If it is assumed that the integrator gain has been chosen so that the amplitude of the signal at the modulator's input is equal to the integrator's output swing, then

$$V_{sw} = V_{DD} - nV_{DS,min}, \qquad (B.10)$$

where n is the number of transistors in series at the amplifier's output, and $V_{DS,min}$ is the minimum drain-to-source voltage allowed in these transistors. To constrain the variation in output resistance with output voltage, $V_{DS,min}$ is typically twice as large as the saturation voltage.

The parasitic capacitance at the amplifier's input is approximately

$$C_P \approx C_{GS1} + C_{w1}, \qquad (B.11)$$

where C_{GS1} is the gate-to-source capacitance of the input transistors, C_{w1} represents wiring parasitics, and the gate-to-drain capacitance of the input transistors has been neglected. If a square-law behavior is assumed for the input transistors, then

$$C_{GS1} = \frac{2L_1^2}{3\mu_P(V_{GS1} - V_{T1})} I_1, \qquad (B.12)$$

where L_1 is the channel length of the input transistors.

To minimize the settling time constant, τ, it was argued in Section 4.3 that C_P should be approximately equal to C_S. Therefore,

$$C_S = C_{GS1} + C_{w1}. \qquad (B.13)$$

Substituting (B.12) into (B.13) and rearranging results in

$$V_{GS1} - V_{T1} = \frac{2L_1^2}{3\mu_P(C_S - C_{w1})} I_1. \qquad (B.14)$$

The maximum value $V_{GS1} - V_{T1}$ is further constrained by dc operating point considerations. Therefore,

$$V_{GS1} - V_{T1} \leq (V_{GS1} - V_{T1})_{max}. \tag{B.15}$$

Finally, $V_{GS1} - V_{T1}$ is also constrained to be greater than some minimum value to avoid operation in the subthreshold region and to ensure that the mismatch between the input transistors does not become too large. Thus,

$$V_{GS1} - V_{T1} \geq (V_{GS1} - V_{T1})_{min}. \tag{B.16}$$

To compute the current consumption in the input differential pair, the following iterative procedure is adopted.

1. C_S is computed using (A.2).
2. V_{sw} is computed using (B.10).
3. An arbitrary value is chosen for I_1.
4. $V_{GS1} - V_{T1}$ is computed using (B.14).
5. The constraints given in (B.15) and (B.16) are verified, and $V_{GS1} - V_{T1}$ is modified if necessary.
6. I_1 is computed using (B.9). If this value is consistent with the previous choice of I_1, then the iteration is complete. Otherwise, a new guess is made for I_1 and steps 4-6 are repeated.

The above procedure reveals the relationship between power dissipation in the input differential pair of an operational amplifier and key technology, circuit, and modulator architecture parameters. This procedure must be combined with those described in Sections B.1, B.2, and B.3 to obtain the power dissipation in the full operational amplifier circuits.

B.1 Folded Cascode Amplifier

The quiescent power dissipated in the folded cascode operational amplifier shown in Figure 4.6(a) is

$$P = (I_1 + 2I_2)V_{DD}. \tag{B.17}$$

Folded Cascode Amplifier

In order to equalize the positive and negative slew rates in the two halves of the differential circuit, M_3 and M_4 must be able to sink a current that is equal to or greater than I_1. Therefore,

$$I_2 \geq \frac{I_1}{2}. \tag{B.18}$$

To ensure stability, the reciprocal of the lowest non-dominant pole in an operational amplifier, τ_2, must be smaller than the settling time constant by some factor, r. Therefore, for the circuit in Figure 4.6(a) it follows from (B.7) with $C_L = 0$ that

$$\tau_2 \leq \frac{(C_S + C_P)(V_{GS1} - V_{T1})}{rI_1}. \tag{B.19}$$

τ_2 is determined from the circuit to be

$$\tau_2 = \frac{C_5}{g_{m5}}, \tag{B.20}$$

where C_5 is the parasitic capacitance at the source of M_5 and g_{m5} is the transconductance of M_5. Substituting (A.24) into (B.20) results in

$$\tau_2 = \frac{C_5(V_{GS5} - V_{T5})}{2I_2}, \tag{B.21}$$

where $V_{GS5} - V_{T5}$ is the overdrive voltage of M_5 and is typically determined by output swing requirements. C_5 is given by

$$C_5 = C_{GS5} + C_{DB1} + C_{DB3} + C_{w2}, \tag{B.22}$$

where C_{GS5} is the gate-to-source parasitic of M_5, C_{DB1} and C_{DB3} are the drain-to-bulk parasitics of M_1 and M_3, respectively, and C_{w2} represents the wiring parasitics at the source of M_5. C_{GS5} is given by

$$C_{GS5} = \frac{2L_5^2}{3\mu_N(V_{GS5} - V_{T5})} I_2. \tag{B.23}$$

The drain-to-bulk capacitances are given by

Power Dissipation vs. Supply Voltage and Oversampling Ratio

$$C_{DB} = W \times C_{DBT}, \tag{B.24}$$

where C_{DBT} is a technology and bias dependent capacitance per unit width, and W is the channel width of the transistor. W is related to the operating point by

$$W = \frac{2IL}{\mu C_{ox}(V_{GS} - V_{TS})^2}. \tag{B.25}$$

To compute I_2 and the power dissipation, the following iterative procedure is adopted.

1. I_2 is set equal to $I_1/2$.
2. The parasitics of the transistors are computed using (B.23) and (B.24).
3. C_5 is computed using (B.22).
4. τ_2 is computed using (B.21).
5. If the constraint given in (B.19) is satisfied, then the iteration is complete. Otherwise, the value of I_2 is increased and steps 2-5 are repeated.
6. The power dissipation can now be computed using (B.17).

This procedure can be used to generate curves such as those shown in Figures 4.7 and 4.8 for the folded cascode amplifier. These curves illustrate the dependence of power dissipation on supply voltage and oversampling ratio. They are thus useful in evaluating the suitability of an operational amplifier topology for a particular application.

B.2 Two-Stage Class A Amplifier

The static power dissipation in the two-stage class A operational amplifier shown in Figures 4.6(b) is

$$P = (I_1 + 2I_2)V_{DD}. \tag{B.26}$$

To ensure that slew limiting occurs only in the first stage of the circuit, the following constraint must be satisfied

$$\frac{2I_2}{C_C + C_I(C_S + C_P)/(C_I + C_S + C_P)} \geq \frac{I_1}{C_C}. \tag{B.27}$$

Two-Stage Class A Amplifier

Substituting (B.8) into (B.27) results in

$$I_2 \geq I_1. \tag{B.28}$$

To stabilize the two-stage class A operational amplifier, the reciprocal of the lowest non-dominant pole in the operational amplifier, τ_2, must be smaller than the settling time constant by some factor, r, as in the case of the folded cascode amplifier. Thus, it follows from (B.7) with $C_L = 0$ that

$$\tau_2 \leq \frac{(C_S + C_P)(V_{GS1} - V_{T1})}{rI_1}, \tag{B.29}$$

τ_2 may be expressed as

$$\tau_2 = \frac{C_5 + C_I(C_S + C_P)/(C_I + C_S + C_P)}{g_{m5}}, \tag{B.30}$$

where C_5 is the parasitic capacitance to ac ground at the gate of M_5, and g_{m5} is the transconductance of M_5. Substituting (A.24) into (B.29) results in

$$\tau_2 = \frac{(C_5 + C_I(C_S + C_P)/(C_I + C_S + C_P))(V_{GS5} - V_{T5})}{2I_2}, \tag{B.31}$$

where $V_{GS5} - V_{T5}$ is the overdrive voltage of M_5 and is typically determined by output swing requirements. C_5 is given by

$$C_5 = C_{GS5} + C_{DB1} + C_{DB3} + C_{w2}. \tag{B.32}$$

To compute I_2 and the power dissipation, the following iterative procedure is adopted.

1. I_2 is set equal to I_1.
2. The parasitics of the transistors are computed using (B.23) and (B.24).
3. C_5 is computed using (B.33).
4. τ_2 is computed using (B.31).
5. If the constraint given in (B.29) is satisfied, then the iteration is complete. Otherwise, the value of I_2 is increased and steps 2-5 are repeated.
6. The power dissipation can now be computed using (B.26).

The curves shown in Figures 4.7 and 4.8 may be generated for the two-stage class A amplifier using the above procedure.

B.3 Two-Stage Class A/AB Amplifier

The quiescent power dissipation in the two-stage class A/AB operational amplifier shown in Figures 4.6(c) is

$$P = (I_1 + 2I_2 + 2I_3)V_{DD} .\tag{B.33}$$

Due to the class AB operation of the second stage, slew limiting only occurs in the first stage of this amplifier. However, unlike the amplifier topologies examined in Sections B.1 and B.2, this circuit has two significant non-dominant poles. For stability considerations, the reciprocals of both of these poles, τ_2 and τ_3, should be smaller than the settling time constant by some factors, r_2 and r_3, respectively. Using (B.7) with $C_L = 0$

$$\tau_2 \leq \frac{(C_S + C_P)(V_{GS1} - V_{T1})}{r_2 I_1},\tag{B.34}$$

$$\tau_3 \leq \frac{(C_S + C_P)(V_{GS1} - V_{T1})}{r_3 I_1}.\tag{B.35}$$

The relationship between τ_2 and the parameters of the circuit is

$$\tau_2 = \frac{C_{11} + C_I(C_S + C_P)/(C_I + C_S + C_P)}{2g_{m11}},\tag{B.36}$$

where C_{11} is the parasitic capacitance at the gate of M_{11} and g_{m11} is the transconductance of M_{11}. Substituting (A.24) into (B.36) results in

$$\tau_2 = \frac{(C_{11} + C_I(C_S + C_P)/(C_I + C_S + C_P))(V_{GS11} - V_{T11})}{4I_2},\tag{B.37}$$

where $V_{GS11} - V_{T11}$ is the overdrive voltage of M_{11} and is typically determined by output swing requirements. C_{11} is given by

Two-Stage Class A/AB Amplifier

$$C_{11} = C_{GS5} + C_{GS11} + C_{DB1} + C_{DB3} + C_{w2}. \tag{B.38}$$

τ_3 is given by

$$\tau_3 = \frac{C_7}{g_{m7}} = \frac{C_7(V_{GS7} - V_{T7})}{2I_3}. \tag{B.39}$$

where C_7 is the parasitic capacitance at the gate of M_7, and g_{m7} is the transconductance of M_7. C_7 is given by

$$C_7 = C_{GS7} + C_{GS9} + C_{DB5} + C_{DB7} + C_{w3}. \tag{B.40}$$

C_{GS7} and C_{GS9} are related through

$$C_{GS7} = \frac{I_2}{I_3} C_{GS7}. \tag{B.41}$$

Substituting (B.41) into (B.40) results in

$$C_7 = C_{GS7}\left(1 + \frac{I_2}{I_3}\right) + C_{DB5} + C_{DB7} + C_{w3}. \tag{B.42}$$

To compute I_2, I_3 and the power dissipation of the two-stage class A/AB amplifier, the following procedure is adopted.

1. I_2 and I_3 are set equal to each other and a value is chosen arbitrarily.
2. C_{11} is computed using (B.38).
3. τ_2 is computed using (B.37).
4. I_2 is adjusted appropriately and steps 2-4 are repeated until the inequality in (B.34) is just satisfied.
5. C_7 is computed using (B.42).
6. τ_3 is computed using (B.39).
7. I_3 is adjusted appropriately and steps 5-7 are repeated until the inequality in (B.35) is just satisfied.
8. Steps 2-8 are repeated until (B.34) and (B.35) are satisfied.
9. The power dissipation can now be computed using (B.33).

Power Dissipation vs. Supply Voltage and Oversampling Ratio

Figures 4.7 and 4.8 plot power dissipation as a function of supply voltage and oversampling ratio, respectively, for a folded cascode amplifier, a two-stage class A amplifier, and a two-stage class A/AB amplifier using the above procedures. The comparatively large output swing and the class AB operation of the second stage make the two-stage class A/AB amplifier the circuit of choice for low to medium frequency applications. However, when the oversampling ratio is high, the relatively high non-dominant pole of the folded cascode amplifier make its power dissipation the lowest among the circuits considered.

APPENDIX C Effects of Capacitor Mismatch

This appendix presents a derivation of the relationship given in (5.13) between the variance of the error mixing coefficient, β, in the 2-1 cascaded modulator of Figure 5.1 and the variance in the values of capacitors used in the implementation of Chapter 6. The error mixing coefficient is realized as a function of the integrator gains shown in Figure 5.9, as given in (5.8). As indicated in (5.12), β can be expressed in terms of capacitances in the integrator circuits of Figures 6.3, 6.4 and 6.5 as

$$\beta = \frac{C_{Sb1}}{C_{I1}} \times \frac{C_{Sa2}}{C_{I2}} \times \left(1 + \frac{C_{Saa3}}{C_{Sb3}}\right), \tag{C.1}$$

where C_{Sai} and C_{Sbi} are the sampling capacitors for the a and b coefficients in the ith integrator, C_{Ii} is the integration capacitor in the ith integrator, and C_{Saa3} represents that portion of the sampling capacitor for a_3 that is not shared with b_3.

To evaluate the impact of capacitor mismatch on β, each of the capacitances is treated as an independent, Gaussian random variable. The calculations can be simplified by noting that if w and x are independent Gaussian random variables with

$$E[w]^2 \gg \sigma[w]^2, \tag{C.2}$$

$$E[x]^2 \gg \sigma[x]^2, \tag{C.3}$$

Effects of Capacitor Mismatch

where $E[\cdot]$ and $\sigma[\cdot]^2$ denote the expectation and variance operators, and if

$$y = \frac{1}{w}, \qquad \text{(C.4)}$$

$$z = w \times x, \qquad \text{(C.5)}$$

then y and z are approximately Gaussian with

$$\sigma[y]^2 \approx \frac{\sigma[w]^2}{E[w]^4}, \qquad \text{(C.6)}$$

$$\sigma[z]^2 \approx \sigma[x]^2 E[w]^2 + \sigma[w]^2 E[x]^2. \qquad \text{(C.7)}$$

Introducing the approximations in (C.6) and (C.7) into (C.1) and simplifying results in

$$\frac{\sigma[\beta]^2}{E[\beta]^2} = \frac{\sigma[C_{Sb1}]^2}{E[C_{Sb1}]^2} + \frac{\sigma[C_{I1}]^2}{E[C_{I1}]^2} + \frac{\sigma[C_{Sa2}]^2}{E[C_{Sa2}]^2} + \frac{\sigma[C_{I2}]^2}{E[C_{I2}]^2}$$
$$+ \frac{\sigma[C_{Saa3}]^2}{(E[C_{Saa3}] + E[C_{Sb3}])^2} + \frac{\sigma[C_{Sb3}]^2 E[C_{Saa3}]^2}{E[C_{Sb3}]^2 (E[C_{Saa3}] + E[C_{Sb3}])^2}. \qquad \text{(C.8)}$$

The variances of the sampling and integration capacitors are related through

$$\sigma[C_{I1}]^2 = b_1 \sigma[C_{Sb1}]^2 \qquad \text{(C.9)}$$

$$\sigma[C_{I2}]^2 = a_2 \sigma[C_{Sa2}]^2 \qquad \text{(C.10)}$$

$$\sigma[C_{Saa3}]^2 = \frac{a_3}{a_3 - b_3} \sigma[C_{Sa3}]^2 \qquad \text{(C.11)}$$

$$\sigma[C_{Sb3}]^2 = \frac{a_3}{b_3} \sigma[C_{Sa3}]^2. \qquad \text{(C.12)}$$

Substituting (C.9)-(C.12) into (C.8) results in

$$\frac{\sigma[\beta]^2}{E[\beta]^2} = \frac{\sigma[C_{Sb1}]^2}{E[C_{Sb1}]^2}(1+b_1^3) + \frac{\sigma[C_{Sa2}]^2}{E[C_{Sa2}]^2}(1+a_2^3)$$
$$+ \frac{\sigma[C_{Sa3}]^2}{E[C_{Sa3}]^2}\left\{\frac{b_3}{a_3-b_3}\left(1+\left(\frac{a_3-b_3}{b_3}\right)^3\right)\right\}.$$

(C.13)

APPENDIX D *Test Setup*

Figure D.1 depicts the test setup used to assess the performance of the experimental ΣΔ modulator described in this work. A list of the instruments used is given in Table D.1. The input to the device under test (DUT) is provided by a high-performance audio signal generator [D.1]. The supply voltages for both the analog and digital portions of the DUT are generated with a dedicated low-noise supply board that is driven by an isolated dc power supply. The two output bits streams of the DUT are fed to a serial-to-parallel converter, and then stored in the digital analysis system (DAS). The data stored in the DAS is subsequently downloaded to a UNIX workstation for processing. The system clocks are generated with a pulse generator that is locked to a precision RF signal source. Details of the various components of the test system are described below.

The modulator's input is generated by a low-distortion balanced sinewave generator. The balanced output and common lines of this signal source are floating with respect to earth ground and are connected to the DUT board though a three conductor shielded audio cable with a 100% foil shield (Belden 8771). The common line is connected to the input common-mode voltage V_{cmi}, and the shield is connected to the DUT board ground. The output of the sinewave generator is terminated on the DUT board with a one-pole low pass filter as shown in Figure D.2. This filter has two functions. First, it bandlimits the input to prevent the aliasing of high-frequency noise into the baseband. Second, it attenuates charge kickback from the sampling switches in the DUT into the source. The loading imposed on the source by this filter can lead to harmonic distortion if the capacitors in the filter are too large. Polystyrene capacitors are

Test Setup

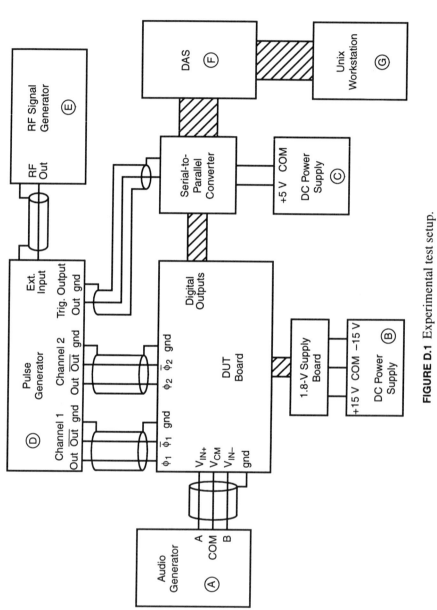

FIGURE D.1 Experimental test setup.

TABLE D.1 Test setup equipment list.

A:	Panasonic VP-7214A RC Oscillator
B:	Hewlett-Packard 6237B Triple Output Power Supply
C:	Hewlett-Packard 6214A Power Supply
D:	Hewlett-Packard 8130A 300 MHz Pulse Generator
E:	Hewlett-Packard 8640B Signal Generator
F:	Tektronix DAS 9200 Digital Analysis System with 92A90 Retargetable Buffer Probe
G:	Digital Equipment Corporation DECstation 3100 with IOTech SCSI 488/D Bus Controller

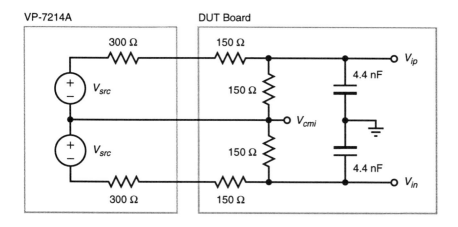

FIGURE D.2 Differential sinewave generator termination circuit.

used since they have a very small voltage coefficient. However, due to their large lead inductance, their self-resonant frequency is relatively low. In higher frequency applications, linear surface mount capacitors (NP0) should be used.

The DUT and supply boards are constructed from two-sided printed circuit boards with an FR4 dielectric layer. The ground planes on the top and bottom sides of the boards are connected by means of a large number of vias and copper tape along the edges of the boards. The DUT board is separated into three regions: 1.8-V analog supply and ground, 1.8-V digital supply and ground, and 5-V supply and ground. The

Test Setup

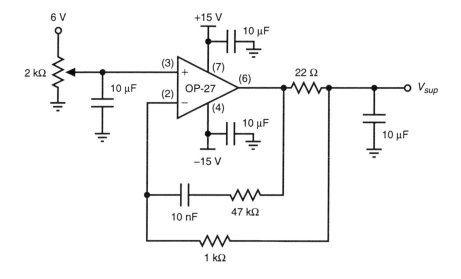

FIGURE D.3 1.8-V power supply.

1.8-V supplies are generated by the supply board and used only in the experimental modulator circuit. The analog supply and ground power the integrators, while the digital supply and ground power the clock generators and the comparator-D/A subsystems. The reference voltages for the D/A converter are applied to the DUT through dedicated pins that are tied to the analog supply and ground on the DUT board. The substrate in the analog circuitry is tied directly to the analog ground, while the substrate in the digital circuitry is tied to a separate substrate ground that is connected to the analog ground on chip at the chip periphery. The DUT is housed in a 44-pin PLCC package with low lead inductance (approximately 10 nH at each pin). The 5-V supply is used by two off-chip comparators on the DUT board, as well as the serial-to-parallel converter board. Two commercial comparators (AD790) convert the low-swing, differential outputs of the DUT to 5-V TTL-level signals that can be read by the serial-to-parallel converter. The three ground planes are separated from each other on the DUT board to avoid injecting switching noise from the digital circuitry into the sensitive analog supply.

The 1.8-V supply board shown as part of Figure D.1 comprises two adjustable low-noise dc voltage sources implemented using OP-27 operational amplifiers, one of which is used for the analog supply and the other for the digital supply. Each of these voltage sources is designed as shown in Figure D.3. The OP-27 acts as a unity gain

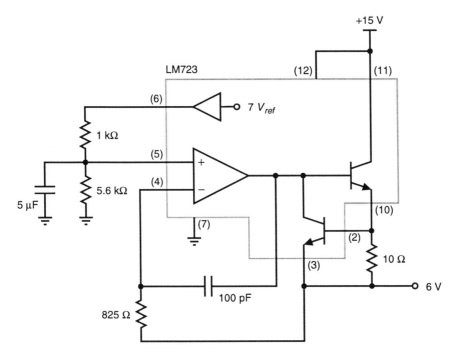

FIGURE D.4 6-V voltage generator.

buffer for the voltage set by the 2-kΩ potentiometer at its input. The supplies are bypassed on the DUT board with the parallel combination of 100-μF, 10-μF, 1-μF, 0.1-μF, and 0.01-μF capacitors, the first two of which are tantalum and the last three ceramic surface mount. The 6-V reference voltage for both of the voltage sources is generated with a single LM723 voltage regulator, as shown in Figure D.4. The supply board is itself powered by a ±15-V power supply that is floating with respect to its earth ground. The common terminal of the power supply is connected to the supply board ground, which is in turn connected to the DUT board ground.

Each of the two bias voltages used in the circuits of Chapter 6, V_{cmi} and V_{mid}, is generated on the DUT board from the 1.8-V analog supply with the circuit shown in Figure D.5. The 10-μF bypass capacitor is tantalum while the other capacitors are ceramic surface mount. The larger capacitors can supply more charge than the smaller ones but have a lower self-resonant frequency. Thus, the capacitor arrangement in

Test Setup

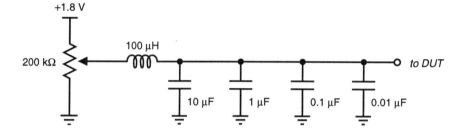

FIGURE D.5 Bias voltage generator.

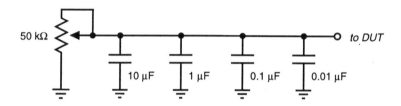

FIGURE D.6 Bias current generator.

Figure D.5 provides decoupling of both low-frequency noise with a large amplitude and high-frequency noise with a small amplitude.

The reference currents for the biasing circuits in the operational amplifiers, shown in Figure 6.11, are generated on the DUT board with the circuit shown in Figure D.6. The bias current lines are bypassed with a capacitor network similar to that used in the bias voltage circuits of Figure D.5.

As discussed in Section 6.3.1, each of the two output bits of the modulator is driven off chip with an open-drain differential pair. The drains of the differential pairs drive 1-kΩ load resistors on the DUT board that are connected to the 1.8-V digital supply. The voltages appearing at the drains are converted to 5-V signals by the comparator circuit shown in Figure D.7. The 1-bit, 4-MHz modulator outputs are converted into 32-bit, 125-kHz output words via serial-to-parallel converters. Each of the two serial-to-parallel converters comprises four TTL 8-bit serial-to-parallel shift registers (SN74164) clocked in parallel into four TTL 8-bit D-type latches (SN74377). The 32-bit outputs thus derived from each modulator output sequence are collected by the

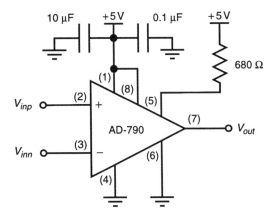

FIGURE D.7 Comparator circuit.

92A90 buffer of the DAS. This buffer is 32 K deep, allowing over one million output bits to be stored for each of the two modulator outputs. Once the modulator's output data has been acquired by the DAS buffer, it is transferred through an IEEE 488 bus to a workstation for subsequent processing.

The modulator and acquisition clocks are generated with a two-channel pulse generator that is locked to a precision RF signal source. The ground of the pulse generator is connected to the DUT board ground and is the only earth ground connection for the entire analog portion of the test setup. The use of the RF signal generator greatly reduces the phase noise in the pulse generator.

The modulator clocks are driven by the output and complementary output of the two channels of the pulse generator. The clocks are connected to the DUT board through SMA cables. Each of the clock lines is terminated to the DUT board ground with a 50-Ω surface mount resistor located near the DUT. A continuous ground plane under the clock lines ensures that the return path for the currents is short. The trigger output of the pulse generator is used to generate the acquisition clock that controls the serial-to-parallel converter and clocks the DAS. The acquisition clock is delayed with respect to the modulator clock to compensate for the propagation delay though the modulator and the output buffer.

Test Setup

REFERENCES

[D.1]   *Panasonic VP-7214A RC Oscillator Instruction Manual*, Matsushita Communication Industrial Co., Ltd., Yokohama, Japan.

Index

A
analog-to-digital conversion 2, 21
  Nyquist-rate 3, 21, 22, 29
  oversampling 2, 3, 21, 22, 29–30
antialias filter 21, 22, 146, 177

B
batteries 1, 3, 13–18, 152
  activation polarization 15
  alkaline 1, 17
  C rate 15
  concentration polarization 15, 16
  discharge profile 17
  E rate 16
  gram-equivalent weight 15
  lithium-ion 18
  nickel-cadmium 1, 14, 17
  nickel-metal-hydride 18
  zinc-carbon 17
bypass network 181, 182

C
chopper stabilization 105
clock booster 137, 152
clock driver 116, 152
clock generator 116, 136
clocks 91, 117, 135, 139
CMOS scaling 1, 3, 8–13, 76, 152
  constant field 9, 10
  constant voltage 9, 10
  generalized 9, 11
common-mode rejection 119
comparator 133, 152
computer-aided design tools
  Matlab 4
  MIDAS 4, 139
correlated double sampling 105, 111, 118

D
data acquisition system 135, 177, 183
decimation filter 3, 4, 23, 50, 57, 75–76, 115, 116, 138–139, 146
delta modulation 31
delta-sigma modulation, see sigma-delta modulation
digital-audio 2, 41, 63, 75, 79, 80, 84, 85, 141
double sampling 136
dynamic latch 133

F
figure of merit, see power dissipation figure of merit 116
fundamental limits 66

185

G
grounding 139, 177, 179–180, 183

H
harmonic distortion 11, 12, 28, 58, 70, 94, 119, 120, 177
hearing aids 152

I
IC package 119, 180
integrator 3, 58, 151
  continuous-time 2, 61–63, 151, 158–159
  leak 108
  settling time 58, 63, 66, 67, 68, 69, 70, 79, 85, 87, 90, 93–95, 105, 106, 107–108, 111, 128, 131, 135, 136, 160, 165, 169, 170
  switched-capacitor 2, 38, 40, 48, 59–61, 70, 88, 90, 93, 97, 101, 151, 155–158
  switched-current 2, 63–66, 151, 155, 159–162
interpolative modulation 32

J
jitter, see timing jitter

K
kickback 133

L
lead inductance 180

M
matching 80
  capacitor matching 79, 85, 92–93, 106–107, 122, 173–175
  transistor matching 12

N
noise
  circuit noise 79, 90, 95–105
  flicker noise 12, 68, 95, 105, 110
  sampling noise 59, 97–103, 105, 109, 110, 119, 120, 122, 129, 131, 155, 159, 160
  switching noise 180
  thermal noise 12, 58, 68, 83, 95, 98, 99, 103–105, 110, 118, 129, 131, 132, 141, 143, 145
noise enhancement factor 95, 104

O
operational amplifier
  common-mode feedback 70, 72–73, 125, 127–128, 130
  common-mode half-circuit 125
  common-mode rejection ratio 115, 143, 151
  dc gain 106, 108, 129, 130, 131
  differential-mode half-circuit 125
  folded cascode 70, 166–168, 172
  ideal class A 60, 62, 155, 156, 158
  ideal class B 61, 62, 146, 155, 157, 159
  power supply rejection ratio 70, 73, 151
  settling time 132
  two-stage class A 70, 168–170, 172
  two-stage class A/AB 70, 72, 75, 76, 124, 143, 151, 152, 170–172

P
phase noise, see timing jitter
power dissipation 151, 155
  figure of merit 3, 116, 146
  fundamental limits 3, 59–66, 68, 155–162

Q
quantization error, see quantization noise
quantization noise 3, 21, 23, 24, 25, 26, 28, 29, 30, 31, 32, 35, 36, 37, 38, 41, 43, 44, 45, 46, 47, 49, 50, 57, 75, 80, 81, 82, 83, 84, 85, 86, 87, 92, 108, 109, 141, 152
quantizer 23, 133
  DAC linearization techniques 49
  DAC nonlinearity 32, 35
  midrise 23
  midtread 23
  performance metrics 28–29
  quantizer gain 23
  uniform 23
quantizer noise model 24–25

R
root-locus method 83

INDEX

S
shielded cable 177
sigma-delta modulation 2, 3, 21, 32, 151
    cascaded 43–49, 79, 80–84, 116, 152
    first-order 21, 33–37
    higher-order 21, 43, 80, 84–85
    integrator leak requirements 47–48, 108
    matching requirements 45–47, 106
    modulator overload 34
    multibit 21, 49–50, 80, 85–87
    noise-differencing modulators 35, 37–43
    sampling noise requirements 109–110
    second-order 38–41, 83, 86
    signal scaling 3, 40, 87–90, 152
    spurious tones 32, 37, 40, 42, 43, 80, 84, 145–146
    stability 43
    thermal noise requirements 110
slew rate 66, 68, 72, 87, 93, 94, 95, 107, 129, 131, 132, 163, 167, 170

T
timing jitter 63, 183

U
unity gain approximation 35, 38, 40, 44, 83, 86

V
voiceband telephony 2, 41, 152
voltage doubler 117, 138, 143

W
white noise approximation 25, 32, 44, 83
wireless applications 152